AF780895

Fusion Bonding/Welding of Polymer Composites

Fusion Bonding/Welding of Polymer Composites

Editors

Patricia Krawczak
André Chateau Akué Asséko
Chung-Hae Park

MDPI • Basel • Beijing • Wuhan • Barcelona • Belgrade • Novi Sad • Cluj • Manchester

Editors

Patricia Krawczak
Centre for Materials
and Processes
IMT Nord Europe,
Institut Mines-Télécom,
University of Lille
Douai, France

André Chateau Akué Asséko
Centre for Materials
and Processes
IMT Nord Europe,
Institut Mines-Télécom,
University of Lille
Douai, France

Chung-Hae Park
Centre for Materials
and Processes
IMT Nord Europe,
Institut Mines-Télécom,
University of Lille
Douai, France

Editorial Office
MDPI
St. Alban-Anlage 66
4052 Basel, Switzerland

This is a reprint of articles from the Special Issue published online in the open access journal *Materials* (ISSN 1996-1944) (available at: https://www.mdpi.com/journal/materials/special_issues/fusion_bonding_welding_polymer_composites).

For citation purposes, cite each article independently as indicated on the article page online and as indicated below:

Lastname, A.A.; Lastname, B.B. Article Title. *Journal Name* **Year**, *Volume Number*, Page Range.

ISBN 978-3-0365-9963-2 (Hbk)
ISBN 978-3-0365-9964-9 (PDF)
doi.org/10.3390/books978-3-0365-9964-9

Cover image courtesy of IMT Nord Europe

© 2024 by the authors. Articles in this book are Open Access and distributed under the Creative Commons Attribution (CC BY) license. The book as a whole is distributed by MDPI under the terms and conditions of the Creative Commons Attribution-NonCommercial-NoDerivs (CC BY-NC-ND) license.

Contents

About the Editors

Patricia Krawczak

Patricia Krawczak graduated with an engineering degree (1989) from Mines Douai, France, then with a MSc (1990) and a PhD in organic and macromolecular chemistry (1993), and finally a diploma of habilitation in physics (1999) from the University of Lille, France. She is currently a Full Professor at IMT Nord Europe, a top-level graduate school of engineering and a research center of the Institut Mines-Télécom (IMT)—the number-one public group of engineering and management graduate schools in France. After having managed IMT Nord Europe's research and education department "Polymers and Composites Technology and Mechanical Engineering" for almost 20 years (2000–2019), she has been in charge of the strategic program "Technology Platforms" of the Institut Mines-Télécom (IMT) since 2019. She has gained deep experience in the field of plastics and composites engineering through numerous collaborations with industry and academia, participated as a principal investigator in numerous national and European industry-oriented projects, and supervised several master's and PhD students. She has co-authored over 400 journal or conference publications and book chapters and is ranked among the World's top 2% of scientists (according to Stanford University/Elsevier data updates). Her research interests cover advanced processing/manufacturing technologies and the physics and mechanics of polymers and polymer composites (including bio-based materials). Additionally, she is a member of the Management Board of the French Society of Plastics Engineers (SFIP) and a member of the Scientific and Strategic Advisory Board of the French Technical Center for Plastics and Composites (IPC). She is also a Knight of the National Order of Academic Palms (decoration for valuable services to the universities in education and teaching) and a Knight of the National Order of the Legion of Honor (highest French order of merit, decoration for public service and professional activity with eminent merits).

André Chateau Akué Asséko

André Chateau Akué Asséko graduated with a MSc degree in thermal and mechanical engineering (2011) from the University of Reims (France) and a PhD degree in optics and lasers (2014) from Mines Douai and University of Lille (France). After two post-doctoral experiences at Mines Albi (France), 2014–2015, and at the French Institute for Bio-based Materials—IFMAS and Mines Douai (France), 2015–2017, he has been appointed as an Associate Professor at IMT Nord Europe (France) since 2017. His research interests include polymer matrix composite materials and their assembly using various processes (such as laser welding and 3D printing) involving multiphysics (i.e., combining thermal, optical, and mechanical aspects) and integrating numerical and experimental approaches. As part of his teaching activities, he also has certified expertise in the design and production of online training content (MOOCs, online courses, and educational videos).

Chung-Hae Park

Chung-Hae Park graduated with BSc and MSc degrees (1996 and 1998) from Seoul National University (Korea), a dual PhD degree (2003) in both mechanical and materials engineering and mechanical and aerospace engineering from Mines Saint-Etienne (France) and Seoul National University (Korea), and holds a diploma of habilitation in Mechanics (2011) from University Le Havre (France). After two experiences in both industry and academia, first as a Senior Research Engineer (2003–2005) at LG Chemical's Technology Centre in Korea and then as a faculty member (2006–2013) at University Le Havre in France, he has been appointed as a Full Professor at IMT Nord Europe (France) since 2013. He has been the deputy director (in charge of innovation) of the Centre for

Materials and Processes of this top-level engineering graduate school since 2021. As a principal investigator or coordinator in numerous national and European industry-oriented projects, he has supervised a number of master's and PhD students. He has co-authored more than 150 journal or conference publications, book chapters, and four patents. His expertise areas cover high-performance structural composite materials (thermoset or thermoplastic matrices; carbon, glass, or vegetal fibers), processing and manufacturing issues (liquid composite molding, resin transfer molding, SMC, compression, and injection molding), with a focus on numerical modeling and simulation.

Preface

Most everyday products and industrial structures are made by assembling several parts. This is the case in the aeronautics, railway, and automotive industries, where polymer composites have become increasingly popular as lightweight substitutes for metallic materials. The joining of polymer composites may be achieved using different technologies. However, one of the most significant drivers for using polymer composites, particularly thermoplastic composites, is the ability to join components via fusion bonding/welding, an attractive alternative to conventional methods, e.g., mechanical fastening and adhesive bonding. Whatever the welding technology, e.g., resistance, induction, ultrasonic, laser, or conduction welding, one of the challenges is to master the interfacial phenomena, structure, and quality in the assembly area (welds), which is rather tricky due to the presence of the reinforcement fibers. The same issues are also tackled for 3D-printed or additively manufactured parts. Additionally, there is a need for reliable predictive process simulation software, as well as in situ monitoring and control of welding process parameters. This Special Issue reprint focuses on the latest advances and development of fusion bonding/welding of polymer composites. It contains research articles addressing materials, processing, modeling, monitoring, or performance issues with either experimental or numerical approaches.

Patricia Krawczak, André Chateau Akué Asséko, and Chung-Hae Park
Editors

materials

MDPI

Article

Radiative Thermal Effects in Large Scale Additive Manufacturing of Polymers: Numerical and Experimental Investigations

Benoît Cosson [1,*], André Chateau Akué Asséko [1], Lukas Pelzer [2] and Christian Hopmann [2]

1 IMT Nord Europe, Institut Mines Télécom, Centre Materials and Processes, University Lille, F-59653 Villeneuve d'Ascq, France; andre.akue.asseko@imt-nord-europe.fr

2 Institute for Plastics Processing (IKV) in Industry and Craft, RWTH Aachen University, Seffenter Weg 201, 52074 Aachen, Germany; lukas.pelzer@ikv.rwth-aachen.de (L.P.); zentrale@ikv.rwth-aachen.de (C.H.)

* Correspondence: benoit.cosson@imt-nord-europe.fr; Tel.: +33-3-27-71-21-93

Abstract: The present paper addresses experimental and numerical investigations of a Large Scale Additive Manufacturing (LSAM) process using polymers. By producing large components without geometrical constraints quickly and economically, LSAM processes have the capability to revolutionize many industries. Accurate prediction and control of the thermal history is key for a successful manufacturing process and for achieving high quality and good mechanical properties of the manufactured part. During the LSAM process, the heat emitted by the nozzle leads to an increase in the temperature of the previously deposited layer, which prepares the surface for better adhesion of the new layer. It is therefore necessary to take into account this part of heat source in the transient heat transfer equation to correctly and completely describe the process and predict the temperature field of the manufactured part. The present study contributes to experimental investigations and numerical analysis during the LSAM process. During the process, two types of measurements are performed: firstly, the heat emitted by the nozzle is measured via a radiative heat sensor; secondly, the temperature field is measured using an infrared camera while varying the process speed. At the same time, a numerical simulation model is developed in order to validate the experimental results. The temperature fields of the manufactured parts computed by numerical simulations are in very good agreement with the temperature fields measured by infrared thermograph with the contribution of the nozzle's heat exchange.

Keywords: additive manufacturing; radiative transfer; large scale; IR measurement

Citation: Cosson, B.; Akué Asséko, A.C.; Pelzer, L.; Hopmann, C. Radiative Thermal Effects in Large Scale Additive Manufacturing of Polymers: Numerical and Experimental Investigations. *Materials* **2022**, *15*, 1052. https://doi.org/10.3390/ma15031052

Academic Editor: Francesca Lionetto

Received: 2 December 2021
Accepted: 27 January 2022
Published: 29 January 2022

Copyright: © 2022 by the authors. Licensee MDPI, Basel, Switzerland. This article is an open access article distributed under the terms and conditions of the Creative Commons Attribution (CC BY) license (https://creativecommons.org/licenses/by/4.0/).

1. Introduction

The industrial use of Additive Manufacturing (AM) is growing rapidly in many areas. The technology's applications range from medical to aerospace. AM allows the manufacturing of parts with complex shapes, difficult to achieve by conventional processes. However, especially for parts obtained by extrusion-based AM technologies such as Fused Deposition Modeling (FDM), mechanical properties of the manufactured parts are oftentimes inferior as compared to parts manufactured by conventional processes such as injection molding. To improve the mechanical properties of parts manufactured by FDM, two approaches can be used. The first one is to improve the intrinsic properties of the constitutive material of the filament by adding reinforcements of different kinds in the thermoplastic [1–4]. The second approach, especially when high mechanical properties are required, is the optimization of process parameters [5–7].

Among the process parameters, the temperature of the extrusion nozzle, as well as its speed of movement, and the temperature of the build plate should be adapted to the mechanical and rheological properties [8]. These parameters directly impact the quality of the manufactured part [9]. An increase of the environmental temperature or the addition of a local material heating system in the FDM process can also have an influence on the manufactured objects [10]. The nozzle temperature (T_N) is one of the parameters having the most effect on the quality and the resistance of the manufactured part [11], since the viscosity of the extruded material depends directly on it [9]. Indeed, by increasing this temperature during the AM process, several researchers [12,13] have observed a drop in viscosity, which then induces transverse flows of the material after deposition, thereby reducing roughness.

The surface quality is improved, as well as the strength of the bond between the different layers obtained by better diffusion through the interfaces. The thermal phenomena (conductive, convective and radiative exchanges) occurring during the process have also been studied by several authors [14,15]. They directly influence the surface quality, dimensional accuracy and mechanical properties of parts. These heat exchanges are among those to be taken into account for modeling and simulating the FDM process [14].

In other cases, the addition of a local laser heat source near the nozzle during extrusion deposition has also been studied by other researchers [16]. During manufacturing, the laser was positioned to heat the strands of the lower substrate layer just before depositing new material on top. The bond strength between layers was found to be increased by 50% and the mode of inter-layer failure became more ductile, with a visible presence of a plastic deformation zone [17].

During the AM process of large-scale parts, the nozzle also emits heat (radiation) on the previously deposited layer which increases in temperature and improves adhesion with the new layer. In one study [18], a desktop FDM machine was used to study the effect of nozzle radiation on the fused filament fabrication process. A proposed 3D numerical model provided information on how the nozzle radiation affects the temperature field of the manufactured part. The temperature fields of the parts computed by numerical simulations were in very good agreement with the temperature fields measured by infrared thermograph.

Based on the mentioned studies, it is evident that temperature conditions influence extrusion-based AM processes and the parts created by using said processes noticeably. To improve process understanding, choice of process parameters as well as predictability and reproducibility, with the goal of obtaining functional parts with high mechanical properties, it is necessary to regard all temperature-related aspects of the AM process. Therefore, this article combines experimental investigations with numerical simulations for a better understanding and improved control of thermal phenomena during extrusion-based AM processes.

2. Experimental Setup

The investigations were carried out using the hybrid manufacturing cell, developed by the institute for plastics processing, Aachen, Germany. It was created to address the major challenges additive manufacturing technologies have to face: production speed, limited part size and limited choice of materials. To improve on state-of-the-art machines, it consists of a screw-based extruder, used for plasticizing and extruding thermoplastic pellets, which is mounted to a six-axis robot arm, providing flexibility in movement and a large build volume. The machine is therefore capable of producing large components in a short production time [19]. The extruder consists of a three-zone screw, mounted inside a divided housing. The intake zone at the upper half of the extruder is water-cooled to prevent premature melting of the material. The lower half is heated by three 100 W heating cuffs which, together with the sheer forces applied by the screw, plasticize the thermoplastic material. The screw is driven by a servo motor coupled to a planetary gearbox. Because of the gear ratio of 320:1, sufficient torque can be applied. To shape the extruded plastic and apply it, layer by layer, to the part being produced, it is pushed through a nozzle with a defined bore. The nozzles used can either be conical or cylindrical in shape, depending on the application.

For example, for processing fiber-filled materials, a cylindrical nozzle geometry would be selected to prevent individual fibers from sticking out in build direction and therefore causing failed parts. On the other hand, if an unfilled material is being processed and the thermal stress on the component should be limited, a conical shape would be advantageous. In combination, the above-mentioned aspects allow for a high material throughput. It is capable of processing thermoplastic pellets, which are less expensive as compared to filament used in other machines. It therefore enables additive manufacturing using standard plastics used in other processes. Furthermore, high levels of filling materials, such as carbon fibers or glass fibers, are possible. Figure 1 shows the extruder and its components.

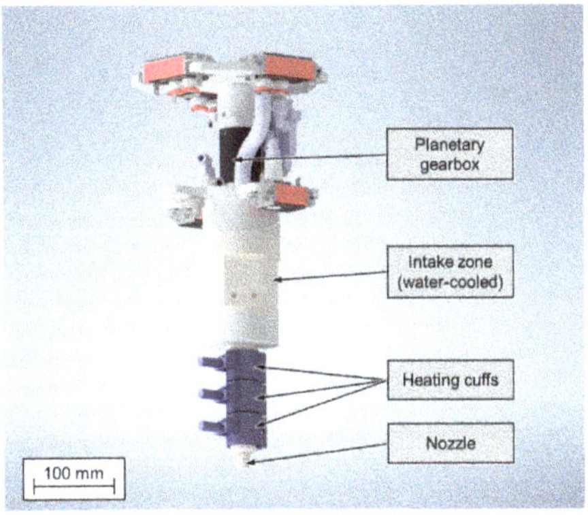

Figure 1. Screw-based extruder for processing thermoplastic pellets in additive manufacturing.

By using the six-axis robot arm, the extruder is moved over the build plate, which is 1900 mm by 600 mm in size. It consists of a 10 mm thick, precision milled aluminum plate which provides flatness across the whole surface. To allow good adhesion between part and build plate, it is covered with a 0.2 mm thin sheet of polypropylene. The sheet is white and semi-translucent. Optionally, the temperature of the joining zone can be influenced locally by either heating or cooling. This enables better layer adhesion as well as better dimensional accuracy of the manufactured component [10].

To measure heat radiation during manufacturing, boxes of 150 mm by 150 mm by 150 mm are produced using a nozzle with a 0.6 mm orifice. The wall thickness is set to be 1 mm and the layer height is configured as 0.3 mm. To assist bed adhesion, a brim of 10 perimeters is used. The boxes are manufactured without infill, bottom or top layers. The material used is the black polypropylene RA130E by manufacturer Borealis AG, Vienna, Austria. The target temperature of the three heaters are, from top to bottom, 265 °C, 280 °C and 335 °C. The in plane displacement velocity of the nozzle is set to 100 $\text{mm}\cdot\text{s}^{-1}$.

Before manufacturing the boxes, calibration of the extruder is needed. This calibration step consist of measuring the emitted radiative heat flux of the extruder and nozzle, as shown in Figure 2a. For the measurement of this radiative heat flux, a radiative sensor by manufacturer Captec, Lille, France, is used. The effective sensor area is 100 mm by 100 mm with a sensitivity of 300 $\text{mV}/(\text{W}/\text{c m}^2)$. The data are recorded directly on a computer.

During this step, the extruder is heated but not filled with material. Therefore, no material is being extruded. The center of the sensor is located at the coordinate $(0, 0, 0)$ and edges of the square sensor are parallel to x and y axes of the machine. The nozzle follows straight lines parallel to the x axis on 9 planes orthogonal to the z axis. In each plane, the nozzle follows 5 equidistant lines. The third line (in the middle) follows the equation $y = 0$. At the middle of this line, the nozzle is directly above the center of the sensor. The aim of this measurement is to isolate and determine the proper radiative heat flux of the extruder during the AM process. In this configuration, the measurements are not perturbed by the heat source for the extruded thermoplastic part that is still in a high temperature state at the beginning of the cooling process. In order to determine the influence of the distance between the sensor and the extruder on the received radiative heat flux, a path of the extruder is designed. This path consist of several round-trips at several heights A numerical simulation of this step is presented in the numerical modeling section and experimental and numerical data will be compared in the results section.

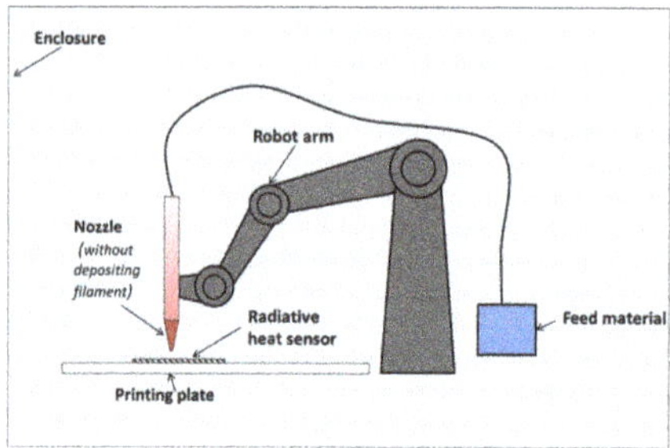

(a)

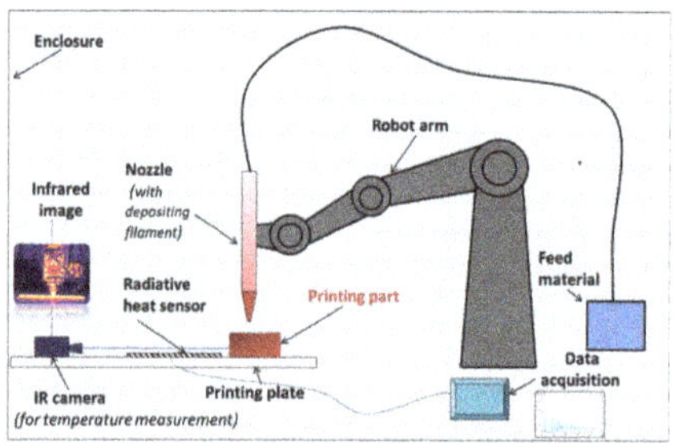

(b)

Figure 2. Experimental setup for determining the amount of thermal radiation during extrusion-based AM processes: (**a**) radiative calibration of the extruder with nozzle over the IR sensor without polymer extrusion, (**b**) infrared measurement from part and extruder during polymer extrusion

A complete monitoring of the AM process is proposed by using the radiative sensor and an IR camera. It is a IR camera (Flir CEDIP JADE) having a spectral response of 3.6 µm–5.1 µm, a sub windowing of 320 pixels × 256 pixels InSb focal plane array (FPA), a 50 mm fixed focus lens and a manufacturer rated precision of ±1%. A computer is connected to the camera for image grabbing and further data analysis with Altair software. This experiment is used to show the effect of self heating of AM parts and the effect of the extruder's radiative heat flux on the temperature field evolution of the part during the process. The box described previously is used in this experiment. The investigations are focused on the temperature evolution of one of the four walls of the box (Figure 3). A simultaneous data recording of the heat flux and the temperature field is done.

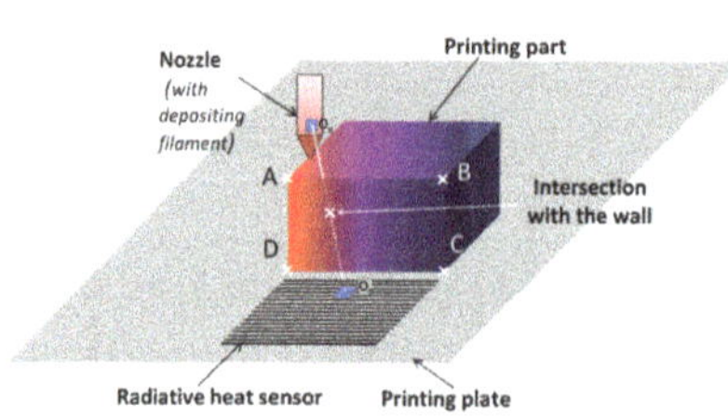

Figure 3. Radiative heat transfer measurement.

3. Numerical Modeling

In order to explain and determine the radiative exchanges, observed and measured during the experimental developments, between the manufactured part and the extruder and between the manufactured part and the environment, a numerical model based on view factors is developed as in [20]. With this model and the data of material properties, nozzle geometry and temperature, a complete description of the LSAM process and the computation of the manufactured part's temperature history is investigated.

For computing the different exchanges, the extruder, the manufactured part and the radiative sensor are discretized in planar surface elements. The surfaces of the extruder, the part and the sensor are discretized in planar face element. By this discretization, it is possible to add up the view factor of each element to the sensor (Equation (1)).

The real shape of the extruder (Figure 1) is very complex and, moreover, the emissivities of the different parts of the extruder are heterogeneous. In order to simplify the computation, the shape of the extruder is defined by a cylinder and a cone. The emissivity and the temperature field are kept constant and homogeneous (Figure 4). O_e is the center of the extruder element, O_s is the center of the sensor element and O is the coordinate system's fixed center. In order to simulate the radiative heat transfer between the extruder and the sensor as well as between the part and the sensor, two numerical strategies are developed. The first case is corresponding to the determination of the radiative power emitted by the nozzle by measuring the heat exchange between the extruder and the sensor when the thermoplastic is not extruded. The geometrical configuration of the experiment, i.e., relative positions of sensor and extruder, is given in Figure 4 [21]. In this configuration, the nozzle path is defined as shown in Figure 5a. There is no obstacle between the extruder and the sensor. The scalar product between the normal vectors $\overrightarrow{N_1}$ and $\overrightarrow{N_2}$ of the two element has to be positive (Equation (2)).

$$dF_{d1-d2} = \frac{\cos(\theta_1)\cos(\theta_2)}{\pi S^2} dA_2 \tag{1}$$

$$\langle \overrightarrow{N_1}, \overrightarrow{N_2} \rangle > 0 \tag{2}$$

In the second case, the thermoplastic is extruded and a square box is manufactured as shown in Figure 3. The geometry of the box is given in Figure 5b. The sensor is placed in front of the manufactured part on the build platform. In the experimental setup, an IR camera records the temperature of the wall that is in the front of the sensor. This wall will have a radiative exchange with the sensor. The temperature of the wall can be in the range from room temperature (fully cooled) to extrusion temperature (newly extruded polymer). In this numerical development, the evolution of the wall's height and temperature are taken into account. The wall's temperature field is extracted from the IR record an then used to compute the radiative exchange between the box and the sensor. During the process, the height of the manufactured walls increase. When the height reaches a certain value, some parts of the extruder can be hidden and cannot be recorded by the sensor when the extruder follows its path around the square section of the manufactured box. For the view factor computation between two elements of the sensor and the extruder, the value is equal to zero when the element of the extruder is hidden by the wall. In order to compute the view factor value, the vector-based geometry is used. The view factor has to be computed for each couple of elements from the extruder and the sensor. If n_e is the number of elements used to discretize the extruder and n_s the number of elements used to discretize the sensor, $n_e \cdot n_s$ view factor computations have to be done to compute the total radiative power received by the sensor from the extruder at each time step. As shown on Figure 3, to know if the intersection between the segment ($[O_eO_s]$), formed by the centers of the two elements (extruder and sensor), and the plane of the wall is inside the rectangle formed by the vertex (named A, B, C, D). In order to compute the intersection between the segment $[O_eO_s]$ and the rectangle ($ABCD$), $ABCD$ is divided in two triangles (ABC and CDA). The intersection is computed for ABC (respectively CDA) as follows:

$$\vec{V} = \frac{\overrightarrow{O_eO_s}}{\|\overrightarrow{O_eO_s}\|} \; ; \; \vec{R} = \overrightarrow{OO_e}$$
$$\vec{X} = [\overrightarrow{OB} - \overrightarrow{OA}, \overrightarrow{OC} - \overrightarrow{OA}, -\vec{V}] \backslash (\vec{R} - \overrightarrow{OA}); \tag{3}$$

$$\begin{cases} u = X(1) \\ v = X(2) \\ t = X(3) \end{cases} \tag{4}$$

$$Test = (u > 0) \;\&\; (v > 0) \;\&\; (u + v > 1) \;\&\; (t > 0)$$

The point of intersection between the segment and the plane is compared with the two triangles. If the point is inside one of the two triangles, the view factor value between the extruder and the sensor is equal to zero. The intersection between ABC respectively CDA and the segment $[O_eO_s]$ occurs when the value of *Test* is *TRUE* (Equation (4)).

In order to compare experimental data to numerical results, the computation of the power ($P_{n \to s}$ Equation (5)) received by the sensor from the nozzle is done for discretized locations on the nozzle path with a virtual displacement speed. The nozzle's value of emissivity ε_n and the part's value of emissivity ε_w are taken equal to 0.95 [18].

$$P_{n \to s} = \frac{1}{S_s} \int_{S_s} \int_{S_n} \varepsilon_n \sigma (T_n^4 - T_s^4) \mathrm{d}F \mathrm{d}S$$
$$P_{w \to s} = \frac{1}{S_s} \int_{S_s} \int_{S_w} \varepsilon_n \sigma (T_w^4 - T_s^4) \mathrm{d}F \mathrm{d}S \tag{5}$$
$$\sigma \approx 5.67 \times 10^{-8}\ \mathrm{W} \cdot \mathrm{m}^{-2} \cdot \mathrm{K}^{-4}$$

The view factor computation depends only on the geometrical configuration in the study of the effect of the extruder on the sensor, the extruder is assumed to have a time independent temperature. This leads to a time independent emitted power. However, for the case of manufacturing the cubic box, where the part's temperature decreases during the cooling process, the radiative power emitted by the box ($P_{w \to s}$ Equation (5)), that is a function of the temperature and the view factor, is changing with time. Moreover, the size of the part changes during the AM process (Figure 6b). In this case, the temperature field (recorded by IR camera) is heterogeneous and time dependent (Figure 6a).

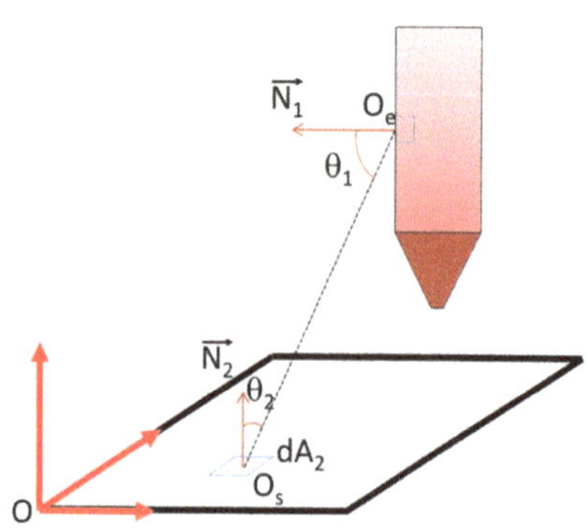

Figure 4. Geometrical configuration for the View factor computation.

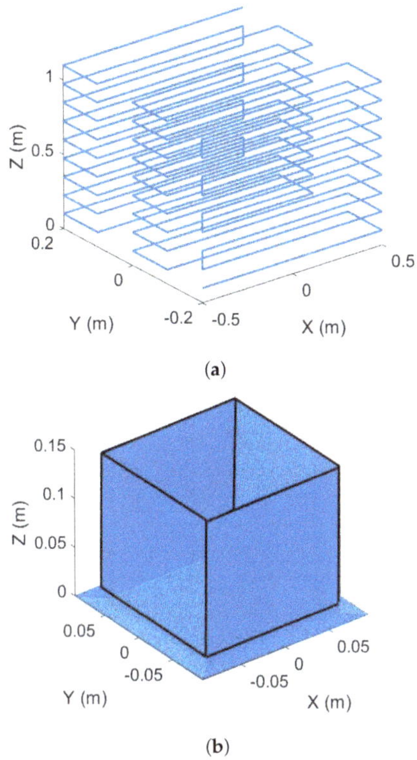

Figure 5. Numerical path of the extruder: (**a**) Path of the extruder over the sensor, (**b**) Path of the extruder to manufacture the part.

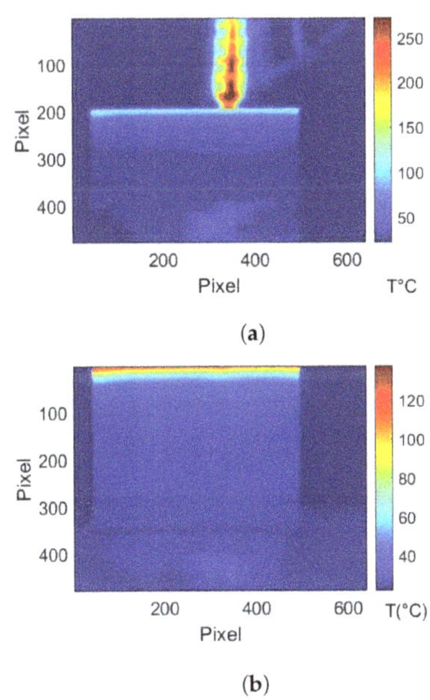

Figure 6. Thermography of the manufactured box: (**a**) at 50% of the process, (**b**) at the end of the process.

4. Results and Discussion

The results of the calibration step where the radiative power emitted by the extruder is measured by the IR sensor is plotted in Figure 7. In this configuration, only the heaters are switched on. There is no extrusion of polymer through the nozzle. In both results, experimental and numerical, the effect of the relative position between the sensor and the nozzle is visible. Each local maximum of the curves corresponds to a position of the nozzle on ($x = 0$) (Figure 5a). The global measured power decreases with the height of the nozzle (Z in Figure 5a). When the position is higher then 0.6 m, for a normalized time over 0.5, the radiative power received by the sensor from the extruder starts to be negligible (Figure 7). Normalized time describes the elapsed time since the start of the measurement divided by total time of the measurement. The view factor is proportional to the inverse square distance between the sensor and the extruder (Equation (1)). Therefore, for large parts, the effect of the nozzle on the evolution of its temperature field only has to be taking into account in a zone of interest that has a radius close to 0.6 m centered around the nozzle. Over 0.6 m the radiative effect of the nozzle is negligible. The results given by the view factor computation are in good agreement with the experimental results. In Figure 7, the local minimum of the received (or calculated) power, when the nozzle is not directly above the sensor, is minimal when the nozzle is in the plane at a height of 0.1 m (for a normalized time less than 0.15). This height is the closest height tested in this first study. Between the normalized time 0.15 and 0.6 the local minimum values of the received power increase. After 0.6, those values decrease. In a first reflection it can be suggested that the local minimum values decrease with the height of the nozzle. In fact, when the numerical results are observed, for small distances between the sensor an the nozzle, there is a competition between the product of the two cosines (the value of this product is equal to zero when $Z = 0$), that is very small, and the square of the distance between the two elements nozzle and sensor (Equation (1)). This effect vanishes when the distance between the sensor and the nozzle has reached a certain value (here 0.8 m), and the values of the angles θ_1 and θ_2 are more stable and the growth of the term S^2 mainly influences the variation of the view factor value. For the following experiment, this observation is important to understand. It also shows that it is important to correctly compute the radiative emissions regarding the geometrical model used to describe the extruder and then correctly simulate the AM process [11].

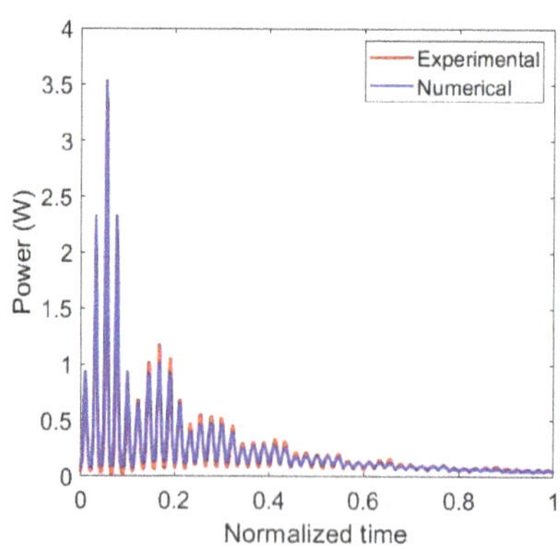

Figure 7. Comparison of experimental and numerical results for the nozzle calibration with the extruder without extrusion of polymer.

Thermographic images where recorded as a video sequence with the infrared camera [22]. The temperature field recording of the manufactured part's front face is done with an OPTRIS IR camera (Figure 6). This camera has a fixed focus lens. A post-treatment of the data is needed to create a continuous crop of the images in order to focus only on the already manufactured part and remove the nozzle and the background. In Figure 6a, the process is at 50 % of the total manufacturing time. In Figure 6b, the thermography is taken at the end of the manufacturing process. These new data are synchronized with the nozzle position and the corresponding view factor computation between the nozzle and the IR sensor. In Figure 6a, the highest wall temperature is recorded at the same level of the extruder temperature [23]. This is the process temperature T_p. The lowest temperature is recorded on the build plate [24]. In Figure 6, the reflection of the radiative emission from the manufactured part can be seen on the build plate. The wall temperature is homogeneous in each horizontal plane. It continuously increases from the build plate to the last deposited layer. The temperature values range from room temperature to the temperature of the extruded polymer. Additionally, the manufactured part is submitted to heat exchanges by convection on every side and by conduction with the build plate.

While manufacturing the box, a brim of ten perimeters is deposited in the first layer, ensuring good adhesion between part build surface during the process (Figure 5b). During this step, the nozzle is close to the build plate and gets closer to the sensor over time (up to 0.05 in normalized time). In Figure 8a, as is not firstly expected, the power received by the sensor decrease while the minimum distance between the sensor and the extruder decrease. This effect is also visible in the numerical results (Figure 8b). This phenomenon is explained previously and it is related to the expression of the view factor (1) and the product of the cosines. In a second time, the value of the power measured by the sensor increase with the height of the manufactured box. The power value reaches its maximum around 0.3 (normalized time) and then decreases till the end of the process. It can be seen that the measured power has a minimum value when the extruder is manufacturing the wall placed at the opposite of the sensor.

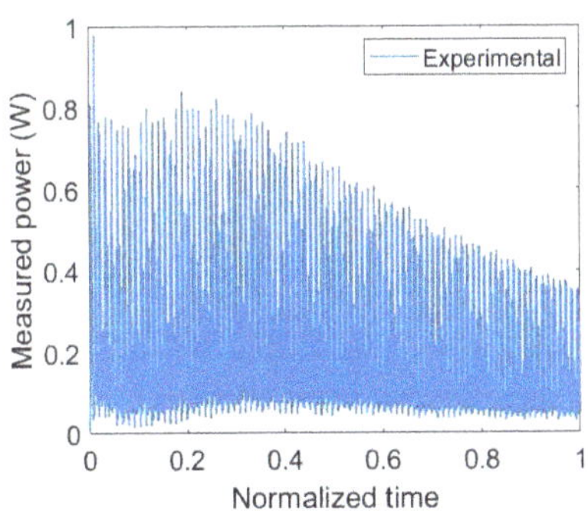

(a)

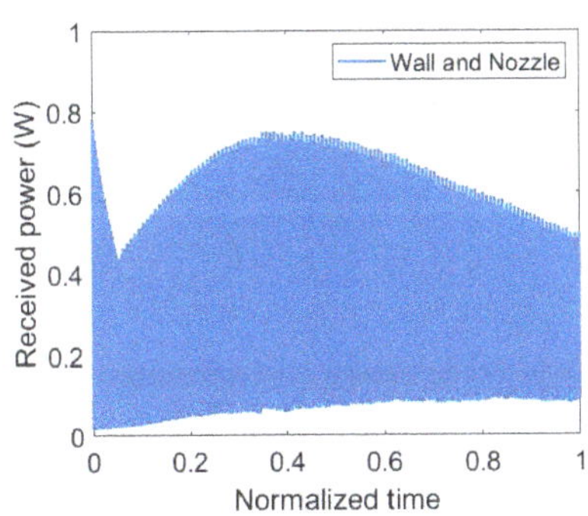

(b)

Figure 8. Radiative heat power on the sensor: (**a**) experimental results, (**b**) numerical results.

In order to explain the different phases of the power evolution, the part due to the nozzle and the part due to the wall of the box are uncoupled and plotted on the Figure 9.

In the first phase, the power is only received from the nozzle. There is only the bottom of the box bonded to the substrate. In the second phase, both of the power received from the nozzle and from the wall are increasing. In the third phase, the power due to the wall still increase to an asymptotic value and the power due to the nozzle decreases. In this phase, when the nozzle is at the opposite of the sensor, the minimum of the power due to the nozzle reaches zero. The height of the box reaches a point where it hides the nozzle from the sensor. The two components of the power, nozzle and wall, have to be simulated to understand the experimental results. The power received from the wall has a monotonic variation contrary to the one received from the nozzle that has an alternative variation [25].

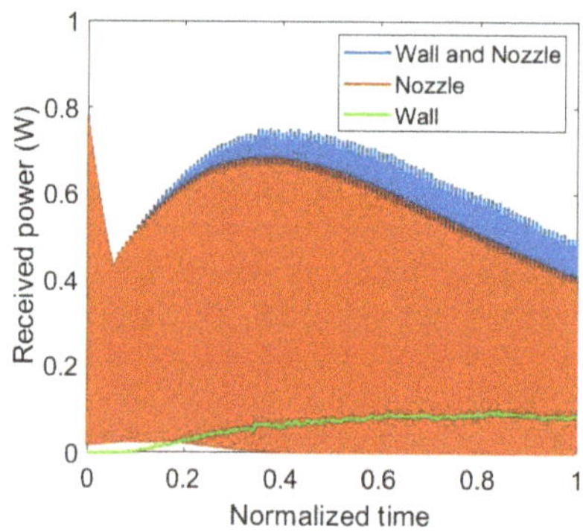

Figure 9. Detailed results for radiative heat power simulation.

5. Conclusions

In this study, developments have been done in numerical and experimental investigations on thermal exchange between the extruder and the printed part during large scale 3D printing. Several new features are presented:

- A numerical model coupled with experimental data was developed for the LSAM polymer process.
- Highlighting of radiative thermal exchanges that should be used for the simulation and optimization of extrusion-based AM processes.
- Confirmation of the importance of the infrared radiation emitted by the nozzle. This radiative energy can help to weld deposited strands on previously manufactured strands by increasing the interface temperature.
- By the numerical results, the radiative power emitted by the wall of the manufactured box is also highlighted.

As shown by the results given in this study, the self-heating during the production process of a part with a hollow shape should be taken into account for numerical simulation of the AM process. It is also shown that the radiative power emitted by the wall is not negligible. In future works, it should be necessary to add all this radiative thermal exchange in a numerical simulation of the material flow during the process in order to correctly compute the bonding between two adjacent strands.

Author Contributions: Conceptualization: B.C. and A.C.A.A.; Numerical works: B.C.; Experimental works: A.C.A.A., B.C., L.P. and C.H.; Writing: B.C., A.C.A.A., L.P. and C.H. All authors have read and agreed to the published version of the manuscript.

Funding: This research was funded in 2019 by I-SITE http://www.isite-ulne.fr/index.php/fr/page-daccueil/; Funded by the Deutsche Forschungsgemeinschaft (DFG, German Research Foundation) under Germany's Excellence Strategy–EXC-2023 Internet of Production–390621612.

Data Availability Statement: The data presented in this study are available on request from the corresponding author.

Conflicts of Interest: The authors declare no conflict of interest.

References

1. Carneiro, O.S.; Silva, A.F.; Gomes, R. Fused deposition modeling with polypropylene. *Mater. Des.* **2015**, *83*, 768–776. [CrossRef]
2. Zhong, W.; Li, F.; Zhang, Z.; Song, L.; Li, Z. Short fiber reinforced composites for fused deposition modeling. *Mater. Sci. Eng. A* **2001**, *301*, 125–130. [CrossRef]
3. Tekinalp, H.L.; Kunc, V.; Velez-Garcia, G.M.; Duty, C.E.; Love, L.J.; Naskar, A.K.; Blue, C.A.; Ozcan, S. Highly oriented carbon fiber–polymer composites via additive manufacturing. *Compos. Sci. Technol.* **2014**, *105*, 144–150. [CrossRef]
4. Melenka, G.W.; Cheung, B.K.O.; Schofield, J.S.; Dawson, M.R.; Carey, J.P. Evaluation and prediction of the tensile properties of continuous fiber-reinforced 3D printed structures. *Compos. Struct.* **2016**, *153*, 866–875. [CrossRef]
5. Christiyan, K.G.J.; Chandrasekhar, U.; Venkateswarlu, K. Influence of raster orientation and layer thickness on mechanical properties of ABS material using FDM process. *Int. J. Adv. Res. Sci. Eng.* **2014**, *3*, 6.
6. Savvakis, K.; Petousis, M.; Vairis, A.; Vidakis, N.; Bikmeyev, A.T. Experimental Determination of the Tensile Strength of Fused Deposition Modeling Parts. In Proceedings of the American Society of Mechanical Engineers Digital Collection, Montreal, QC, Canada, 14–20 November 2014. [CrossRef]
7. Shofner, M.L.; Lozano, K.; Rodríguez-Macías, F.J.; Barrera, E.V. Nanofiber-reinforced polymers prepared by fused deposition modeling. *J. Appl. Polym. Sci.* **2003**, *89*, 3081–3090. [CrossRef]
8. Le, A.D.; Cosson, B.; Asséko, A.C.A. Simulation of large-scale additive manufacturing process with a single-phase level set method: A process parameters study. *Int. J. Adv. Manuf. Technol.* **2021**, *113*, 3343–3360. [CrossRef]
9. Hwang, S. Study of Materials and Machines for 3D Printed Large-scale, Flexible Electronic Structures Using Fused Deposition Modeling. Available online: https://scholarworks.utep.edu/open_etd/1072/ (accessed on 26 January 2022).
10. Hopmann, C.; Dahlmann, R.; Lammert, N.; Pelzer, L.; Hellmich, C. Process and design optimisation in additive manufacturing. In Proceedings of the 30th International Colloquium Plastics Technology, Online, 8–11 September 2020; Shaker Verlag: Düren, Germany, 2020; ISBN 978-3-8440-6892-4.
11. Thézé, A.; Régnier, G.; Guinault, A.; Richard, S.; Macquaire, B. Fused filament fabrication printing process of polymers highly filled with metallic powder: A significant influence of the nozzle radiation on the substrate temperature. *Int. J. Mater. Form.* **2021**, *14*, 1511–1521. [CrossRef]
12. Hwang, S.; Reyes, E.I.; Moon, K.s.; Rumpf, R.C.; Kim, N.S. Thermo-mechanical Characterization of Metal/Polymer Composite Filaments and Printing Parameter Study for Fused Deposition Modeling in the 3D Printing Process. *J. Electron. Mater.* **2015**, *44*, 771–777. [CrossRef]
13. Reddy, B.V.; Reddy, N.V.; Ghosh, A. Fused deposition modelling using direct extrusion. *Virtual Phys. Prototyp.* **2007**, *2*, 51–60. [CrossRef]
14. Costa, S.; Duarte, F.; Covas, J. Thermal conditions affecting heat transfer in FDM/FFE: A contribution towards the numerical modelling of the process. *Virtual Phys. Prototyp.* **2015**, *10*, 35–46. [CrossRef]
15. Sun, Q.; Rizvi, G.; Bellehumeur, C.; Gu, P. Effect of processing conditions on the bonding quality of FDM polymer filaments. *Rapid Prototyp. J.* **2008**, *14*, 72–80. [CrossRef]
16. Du, J.; Wei, Z.; Wang, X.; Wang, J.; Chen, Z. An improved fused deposition modeling process for forming large-size thin-walled parts. *J. Mater. Process. Technol.* **2016**, *234*, 332–341. [CrossRef]
17. Ravi, A.K.; Deshpande, A.; Hsu, K.H. An in-process laser localized pre-deposition heating approach to inter-layer bond strengthening in extrusion based polymer additive manufacturing. *J. Manuf. Process.* **2016**, *24*, 179–185. [CrossRef]
18. Cosson, B.; Akué Asséko, A.C. Effect of the Nozzle Radiation on the Fused Filament Fabrication Process: Three-Dimensional Numerical Simulations and Experimental Investigation. *J. Heat Transf.* **2019**, *141*, 082102. [CrossRef]
19. Hopmann, C.; Pelzer, L. Additive Manufacturing of Large Parts by an Extrusion based Hybrid Approach. In Proceedings of the Polymer Processing Society–PPS Conference Proceedings, Cesme-Izmir, Turkey, 26–30 May 2019.
20. Cosson, B.; Schmidt, F.; Maoult, Y.; Bordival, M. Infrared heating stage simulation of semi-transparent media (PET) using ray tracing method. *Int. J. Mater. Form.* **2011**, *4*, 1–10. [CrossRef]
21. Ray, M.S. *Thermal Radiation Heat Transfer*, 3rd ed.; Siegel, R., Howell, J.R., Eds.; Hemisphere Publishing Corporation: New York, NY, USA, 1992; 1072p, ISBN 0-89116-271-2. [CrossRef]
22. Seppala, J.E.; Migler, K.D. Infrared thermography of welding zones produced by polymer extrusion additive manufacturing. *Addit. Manuf.* **2016**, *12*, 71–76. [CrossRef] [PubMed]
23. Lepoivre, A.; Boyard, N.; Levy, A.; Sobotka, V. Heat Transfer and Adhesion Study for the FFF Additive Manufacturing Process. *Procedia Manuf.* **2020**, *47*, 948–955. [CrossRef]
24. Compton, B.G.; Post, B.K.; Duty, C.E.; Love, L.; Kunc, V. Thermal analysis of additive manufacturing of large-scale thermoplastic polymer composites. *Addit. Manuf.* **2017**, *17*, 77–86. [CrossRef]
25. Wolszczak, P.; Lygas, K.; Paszko, M.; Wach, R.A. Heat distribution in material during fused deposition modelling. *Rapid Prototyp. J.* **2018**, *24*, 615–622. [CrossRef]

materials

Article

Assembling of Carbon Fibre/PEEK Composites: Comparison of Ultrasonic, Induction, and Transmission Laser Welding

Adrian Korycki [1], Christian Garnier [1], Margot Bonmatin [1,2], Elisabeth Laurent [3] and France Chabert [1,*]

1 Laboratoire Génie de Production, ENIT-INPT, University of Toulouse, 47 Avenue d'Azereix, 65016 Tarbes, France
2 Institut Clement Ader (ICA), University of Toulouse, CNRS, IMT Mines Albi, INSA, ISAE-SUPAERO, UPS, Campus Jarlard, 81013 Albi, France
3 CNES, Sous-Direction Assurance Qualité, Service Technologies, Matériaux et Procédés, 18 Avenue Edouard Belin, 31401 Toulouse, France
* Correspondence: france.chabert@enit.fr

Abstract: In the present work, an ultrasonic, an induction, and a through transmission laser welding were compared to join carbon fibre reinforced polyetheretherketone (CF/PEEK) composites. The advantages and drawbacks of each process are discussed, as well as the material properties required to fit each process. CF/PEEK plates were consolidated at 395 °C with an unidirectional sequence and cross-stacking ply orientation. In some configurations, a polyetherimide (PEI) layer or substrate was used. The thermal, mechanical, and optical properties of the materials were measured to highlight the specific properties required for each process. The drying conditions were defined as 150 °C during at least 8 h for PEI and 24 h for CF/PEEK to avoid defects due to water. The optical transmission factor of PEI is above 40% which makes it suitable for through transmission laser welding. The thermal conductivity of CF/PEEK is at most 55 $W\cdot(m\cdot K)^{-1}$, which allows it to weld by induction without a metallic susceptor. Ultrasonic welding is the most versatile process as it does not necessitate any specific properties. Then, the mechanical resistance of the welds was measured by single lap shear. For CF/PEEK on CF/PEEK, the maximum lap shear strength (LSS) of 28.6 MPa was reached for a joint obtained by ultrasonic welding, while an induction one brought 17.6 MPa. The maximum LSS of 15.2 MPa was obtained for PEI on CF/PEEK assemblies by laser welding. Finally, interfacial resistances were correlated to the fracture modes through observations of the fractured surfaces. CF/PEEK on CF/PEEK joints resulted in mixed cohesive/adhesive failure at the interface and within the inner layers of both substrates. This study presents a guideline to select the suitable welding process when assembling composites for the aerospace industry.

Keywords: thermoplastics; composites; laser welding; induction welding; ultrasonic welding

Citation: Korycki, A.; Garnier, C.; Bonmatin, M.; Laurent, E.; Chabert, F. Assembling of Carbon Fibre/PEEK Composites: Comparison of Ultrasonic, Induction, and Transmission Laser Welding. *Materials* **2022**, *15*, 6365. https://doi.org/10.3390/ma15186365

Academic Editor: Francesca Lionetto

Received: 10 August 2022
Accepted: 7 September 2022
Published: 13 September 2022

Copyright: © 2022 by the authors. Licensee MDPI, Basel, Switzerland. This article is an open access article distributed under the terms and conditions of the Creative Commons Attribution (CC BY) license (https://creativecommons.org/licenses/by/4.0/).

1. Introduction

Polymer welding is a process of assembling surfaces of thermoplastic-based materials, generally with heat and under pressure [1–3]. Only thermoplastics of the same nature or miscible with each other can be assembled by welding. Welding is a relevant assembling process to contribute to the reduction of the environmental impact of the lightening of structures. Besides, as polymer welding is fast, safe, and cheap, spreading such a process in the industry will contribute to the competitiveness of companies. Welding thermoplastics is quite well mastered, whereas more research is required to target welding of short and long-fibre thermoplastic composites. Moreover, some issues remain to make the shift toward welding complex shapes and large structures of thermoplastic polymers and composites.

A choice of a process is affected by materials to be joined, a joint configuration, a required mechanical strength, a level of seal, process costs and speed, and production volumes. However, the creation of a good quality weld does not only depend on the welding process but also on the weldability of the materials to be joined [4]. Therefore, the evaluation of their weldability is of high importance in the whole welding operation for a successful assembly of polymer materials.

Compared to thermosets, thermoplastics can be recycled and do not require refrigerated storage, offering almost an infinite shelf life. Most of the thermoplastic composites used in marine, aerospace, and automotive lightweight structures are made of carbon fibre reinforced polymers (CFRP) composites, which give way to glass fibre reinforced polymers (GFRP) for some applications. Carbon fibres give better properties than glass fibres because of their higher stiffness and strength. Sometimes, the carbon fibres used are high modulus (HM) for the highest performance required in some applications. The glass or carbon fibres associated with a matrix of whether polyphenylene sulfide (PPS), polyetherimide (PEI) or polyaryletherketone (PAEK), mostly polyetheretherketone (PEEK) and polyetherketoneketone (PEKK) [3,5]. These high-performance polymers display high damage tolerance to finished parts, as well as chemical resistance, and, thus, are stable in severe hot or wet conditions. Thermoplastic composites can be re-melted with promising benefits in repairing and end-of-life re-using. Thermoplastics are a response to the growing demand for new materials that can be used in aircraft structures with a reduced total weight while maintaining high mechanical properties. In addition, despite the high price of thermoplastics such as polyetherketone (PEK), PEEK, and PPS, the total cost of part manufacturing is lowered as a result of a significant reduction in production times.

Moreover, thermoplastic composites offer the ability to save weight and improve the sustainability of airplanes and spacecrafts by joining components via fusion bonding. Welding is an attractive alternative to conventional methods, such as mechanical fastening and adhesive bonding to join composite parts. However, since composites gain their outstanding properties from the fibres, the weld is inevitably the weakest point in the system as the weld is a zone free of fibres.

The welding processes are mainly focused on a couple of advanced materials, such as carbon fibre reinforced polyetheretherketone. Also, glass and carbon fibre reinforced polyetherimide (GF/PEI and CF/PEI) have been investigated. CF/PEEK is in most cases the APC-2, a semi-product manufactured by Imperial Chemical Industries (UK), while PEI-based materials are known as Cetex, supplied by Ten Cate Advanced Composites (The Netherlands). Some studies were also attracted to polyphenylene sulphide or polypropylene (CF/PPS or GF/PP and CF/PP) [6].

Many welding techniques have been proposed, developed, and evaluated for thermoplastic applications. Based on the mechanism of heat generation at the welding interface, welding methods for thermoplastics can be classified as external (thermal heat source) and internal (mechanical movement or electromagnetism) heating methods. They differ in the way heat is generated at the interface, such as frictional heating (e.g., ultrasonic welding [7–10]), electromagnetic heating (e.g., induction welding [11]), and thermal techniques (e.g., laser welding [12–17]). The selection of the welding method depends mainly on the thermal and electrical properties of the material and the shapes of the elements to be joined [18]. Ultrasonic, induction, and laser welding are the most used processes to assemble such thermoplastic composites.

Whatever the process, the physical mechanisms during welding of thermoplastics are accomplished in three sequential stages: a surface preparation, an application of heat, and a weld creation on cooling. Upon pressure, the surfaces come into close contact which initiates a macromolecular diffusion and an entanglement across the joint. Only thermoplastics and thermoplastic-matrix composites can be welded [19]. Compared to thermosets, their chains are free to move up to long distances when intermolecular weak bonds are broken. All the welding processes are governed by the same parameters: heat, pressure, and time [20]. To achieve high-quality welds, a careful optimization of the welding parameters is required for each application.

In ultrasonic welding, a sonotrode coupled to a transducer that produces high frequency (20–40 kHz) vibration causes frictional heat and melting at the interface of two parts. A polymeric film is required at the interface as an energy director. Among the results published until now, studies on high-performance thermoplastic composites include

CF/PEI [21,22], CF/PPS [23–25], and CF/PEEK [10]. The strength of joints in ultrasonic welding has been modeled and reported by Benatar and Gutowski [26] for PEEK/AS4 graphite APC-2 composites. The strength of the welded parts obtained was estimated as 74 MPa. The effect of welding time was investigated by Tao et al. [10] for CF/PEEK with a flat PEEK film as an energy director. It was reported that with a gradual increase in welding time, the weld strength also increased up to an optimum time of 0.9 s, the maximum LSS obtained was 28 MPa. Longer welding time leads to overheating and local degradation [10]. Recently, CF/PEEK assembled by an ultrasonic welding reached a maximum LSS at 49 MPa [27]. The same authors provided thermal profiles during welding from microthermocouples located close to a welded interface. The evolution of the interfacial temperature during welding was correlated to welding parameters to prevent overheating. Moreover, a continuous ultrasonic welding has been currently under development [7,28,29].

In induction welding, an electromagnetic coil produces alternating electromagnetic fields which induce Eddy currents in conductive laminate. Eddy currents produce heat through carbon fibres or a metallic susceptor which melts the thermoplastic. The perpendicular 0° and 90° fibre orientation in the woven fabric are ideal, enabling Eddy currents to be generated in each ply of the laminate. With unidirectional laminated stacks, however, it is common to have 45° plies interspersed so that the angle difference is smaller. Williams et al. [30] by using a woven fabric susceptor made from co-spun fibre yarn comprising blended staple carbon and PEEK fibres welded APC-2 laminates. An LSS of 46 MPa was obtained. However, induction welding without a susceptor required the electrical and thermal conductivity of the materials to fit the process. The main consideration for design was how to control and concentrate the magnetic field onto the workpiece. Using no susceptor, Cogswell et al. [31] were able to obtain an LSS of 31 MPa for APC-2 laminates joined. Border and Salas [32] investigated induction welding on PEEK without metal susceptors. An LSS of 48.2 MPa was obtained. Reis et al. [3] review presented that the effectiveness of induction welding in several thermoplastic composites was evaluated with lap shear strength values ranging between 14 MPa and 43 MPa.

During the transmission-through laser welding process, laser radiation passes through an upper laser transparent part, for instance, unreinforced amorphous PEEK and it generates heat in a lower laser absorbent part which contains optionally carbon fibre or conductive additive. Amanat et al. [33] welded PEEK films using a pulsed fibre laser (1060 nm) and a maximum power of 20 W. According to their results, the two lowest scan speeds, 4 $mm \cdot s^{-1}$ and 8 $mm \cdot s^{-1}$, showed the most significant bond strengths of around 22 MPa to 25 MPa for semi-crystalline and 12 MPa to 19 MPa for amorphous PEEK. Since carbon fibres are laser absorbent, CF/PEEK is suitable as a lower part only. The reliable application of laser welding technology for joining primary aeronautic parts made of thermoplastic composites still requires significant development and investigation [15]. Torrisi et al. [34] used 3-ns Nd: Yag laser (532 nm) system with various fillers for ultra-high molecular weight polyethylene composites. They found that the best polymer coupling was with carbon nanotubes which showed a good adhesion value with 10 MPa shear rupture at an irradiation time of 2 min.

Ultrasonic, induction, and laser welding processes are already widespread for assembling pure thermoplastics. However, welding fibre-reinforced composites bring new issues to be tackled compared to pure thermoplastics. Indeed, the fibres modify the properties of the materials, mainly optical, thermal conductivity, and rheological properties. As a consequence, all the process parameters need an overhaul to reach repetitive and reliable welds.

The purpose of this study is to compare three welding processes for assembling high-performance thermoplastic composites for spacecraft applications. Whereas each existing study focuses on one process/material couple, the originality of the hereby article is to compare three welding techniques to join the same materials. Ultrasonic welding (UW), induction welding (IW), and transmission laser welding (TLW) are applied to assemble CF/PEEK thermoplastic composites and PEI on CF/PEEK. The materials are characterized to measure their water intake, thermal transitions, thermal stability, thermal conductivity thermomechanical properties, and optical properties. Then, the mechanical resistance of welds is measured by single lap shear tests. Finally, the main strengths and flaws of each process are highlighted as well as the material requirements for successful welding of CF/PEEK composites.

2. Materials and Methods

2.1. Materials and Samples Preparation

The materials used in this study are:

- Thermoplastic composite prepregs carbon fibre HM63/PEEK developed by Suprem (Yverdon-les-Bains, Switzerland) and supplied by Centre National d'Etudes Spatiales (CNES, Toulouse, France) as 140 μm thick tapes. The latter, named CF/PEEK in the following, was made of PEEK 150G from Victrex as a matrix and unsized carbon fibre HM63 from Hexcel. HexTow HM63 carbon fibre is a continuous, high strength of 4.5 GPa and high modulus of 452 GPa, with a density of 1.83 $g \cdot cm^{-3}$, polyacrylonitrile-based carbon fibre.
- PEEK 450 G from Victrex (Lancashire, UK) in granules for DMTA and TGA only.
- PEI Ultem 1000 from Sabic (Rijad, Saudi Arabia) in granules.
- PEI Ultem 1000 as 250 μm thick film was purchased from GoodFellow (Lille, France).

CF/PEEK, pure PEEK, and PEI were processed by compression moulding. A hydraulic press LAB 800P PEI from Pinette Emidecau Industries (Chalon-sur-Saône, France) was used for preparing 1, 2, and 4 mm thick plates. The materials were dried in a vacuum oven at 150 °C for at least 3 h before processing. The plates of CF/PEEK 1 mm with a unidirectional sequence of the tapes $[0°_7]$ (Figure 1a) and plates with cross stacking $[45°,-45°,0°_3,-45°,45°]$ (Figure 1b) were consolidated at 395 °C and 2 MPa for 30 min, the heating, and cooling rates were 4 $°C \cdot min^{-1}$. The unidirectional sequence was expected to dissipate heat faster and more efficiently. This configuration is the worse for fusion bonding because most of the generated heat is dissipated on each side of the specimen. Higher temperatures must be attained to melt the interface to be welded. The cross-stacking sequence is close to industrial cases and it was adapted to fit the induction welding process by adding a 45° layer on the upper surface of the specimens. PEEK pellets were kept in a cavity of 2 mm thickness between two steel foils, and the temperature was increased at a speed of 10 $°C \cdot min^{-1}$ up to 360 °C. A pressure of 10 MPa was applied at 360 °C for 5 min. Then, still under pressure, the moulded samples were cooled down to 200 °C at the speed of 4 $°C \cdot min^{-1}$ before demoulding. PEI plates with a nominal thickness of 1, 2, and 4 mm were produced from granules at 290 °C and 10 MPa for 5 min, and heating and cooling rates were 10 $°C \cdot min^{-1}$. The plates with dimensions of 100×50 mm^2 (1 mm thick), 150×75 mm^2 (2 mm thick), and 110×75 mm^2 (4 mm thick) were cut into samples of 100 mm long and 25 mm wide.

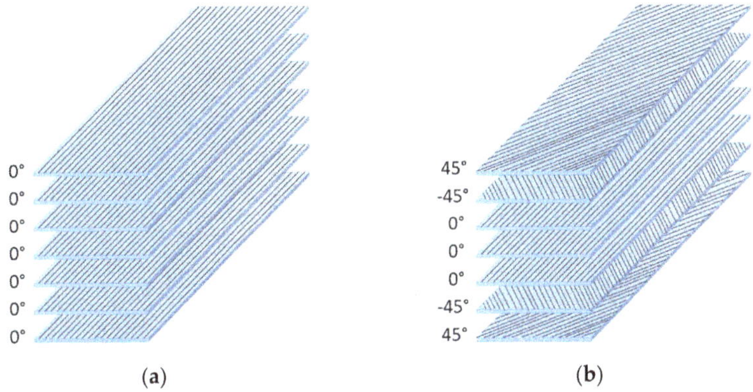

Figure 1. Sequence of the CF/PEEK films (**a**) unidirectional and (**b**) cross stacking.

2.2. Characterization of Materials

During welding, between the two elements to be joined, diffusion and interface crossing of macromolecules take place. The physical phenomena of macromolecular diffusion, healing, and crystallization at the interface are well described in Martineau's work [35] for PEEK.

After cooling, a permanent cohesive joint is formed. The appropriate mobility of the diffusing chains is reached when the temperature is above the melting temperature for semi-crystalline polymers or well above glass transition for amorphous polymers. Also, the materials to be joined should have a similar softening or melting point, must be miscible, or must have a similar chemical structure [36]. Moreover, thermoplastics are sensitive to thermo-oxidative degradation. During welding, overheating can occur and it weakens the interfacial strength [37].

To understand the interactions of the welding parameters with the specimens, careful consideration of the material properties is beneficial. Main properties were identified as playing a role in the welding result and divided into two groups:

(1) roughness and surface chemistry,
(2) rheological, optical, and thermal properties.

For the sake of brevity, only the second group is presented in this article. Moreover, the presence of water within the specimens has been proved to be detrimental to the joint integrity and mechanical resistance of welds [38]. For this reason, careful drying is necessary before assembling the specimens. Also, the water content could influence further material characteristics.

2.2.1. Drying of Specimens

PEEK and PEI are sensitive to water because they contain chemical groups such as ketones, ethers, and hydroxyls which create weak bonds with H_2O. So, a thorough study of the kinetics of sorption and drying was carried out to determine an appropriate drying procedure. Our approach consists of three steps:

(1) drying at 150 °C for at least 3 h,
(2) immersion of the specimens in a beaker containing deionized water at 25 °C. A thermocouple IKA ETS-D5 from KA-WERK (Brachbach, Germany) was placed inside the beaker and the latter was maintained on a hotplate stirrers IKA C-MAG HS7 from KA-WERK (Brachbach, Germany) to keep a constant temperature throughout the experiment,
(3) the specimens were placed in an oven Memmert UNB 200 (Schwabach, Germany) at 150 °C. Mass readings with a Mettler Toledo scale (Viroflay, France) during sorption and drying were taken over a short time interval at first every hour, then twice a day.

Besides the evolution of weight with time, the presence of water inside the materials was checked by Fourier Transform Infrared Spectrometry. Spectra were recorded using an FTIR spectrometer Spectrum One by Perkin Elmer (Waltham, MA, USA) in attenuated total reflectance (ATR) mode in the 4000–650 cm^{-1} range. The resolution was 4 cm^{-1}, and 16 scans were accumulated for an improved signal-to-noise ratio.

2.2.2. Dynamic Mechanical Analysis

Thermomechanical properties were measured by dynamic mechanical thermal analysis (DMTA) using ARES LN2 rheometer from TA Instruments (New Castle, DE, USA) in torsion mode. Parallelepipedic specimens of 45 × 10 × 2 mm^3 were cut out from moulded plates. Temperature ramps from 25 °C to 325 °C at a heating rate of 3 °C·min^{-1} were applied at a frequency of 1 Hz and strain of 0.1% within the linear viscoelastic (LVE) domain of both PEEK and PEI. Previously, strain sweeps were carried out to define the LVE domain.

2.2.3. Optical Properties

Optical properties are very important for transmission laser welding. Transmission factors of PEEK and PEI were measured with a VERTEX 70 spectrophotometer from Bruker (Billerica, MA, USA) in the spectral range between 400 nm and 20.000 nm. The transmission mode with an integrating sphere and an incident angle of 0° between the beam and the normal to the sample was applied. The tests were carried out at room temperature with the integrating sphere. The evolution of the transmittance as a function of temperature was measured between 20 °C and 140 °C with an increment of 20 °C.

2.2.4. Thermal Properties

Only thermoplastics whose processing temperatures are in the same range can be welded together. Both parts must be melted to allow the diffusion of macromolecules through the interface. Thermal transitions of PEEK and PEI were measured by differential scanning calorimetry with a DSC 1 Star System from Mettler Toledo (Viroflay, France). The measurements were performed under a nitrogen flow of 50 mL·min^{-1}, and a mass of approximately 10 mg was placed in sealed aluminium pans Mettler Toledo (Viroflay, France). The DSC scan comprised two steps:

(1) an ascending temperature sweep of 10 °C·min^{-1}, from 25 °C to 400 °C,
(2) a descending temperature sweep of 10 °C·min^{-1}, from 400 °C to 25 °C.

Glass transition temperature (T_g), crystallization temperature (T_x), and melting temperature (T_m) were obtained.

Then, the behavior of materials under high-temperature conditions was evaluated through a TGA 2 Mettler Toledo device (Viroflay, France). The degradation temperature (T_d) was obtained using platinum crucibles Mettler Toledo (Viroflay, France) under an oxygen atmosphere at 50 mL·min^{-1}, sample mass of 10–15 mg, and the heating rate at 5 °C·min^{-1} in the range from 25 °C to 800 °C.

Thermal conductivity and specific heat were measured by using Hot Disk TPS 2500 S analyzer (Göteborg, Sweden). The hot disk probe was placed between two 5 mm thick samples of the material to be characterized. Surfaces of the samples in contact with the probe are flat. The principle is to pass an electric current through the probe to generate a temperature increase of one to several degrees and to record the increase in resistance (temperature) over time. The temperature of the measurement was 23 °C.

2.2.5. Degree of Crystallinity

The mass fraction of the crystalline phase was calculated from the melting enthalpy measured by DSC and the theoretical melting enthalpy of 100% crystalline phase with the following formula:

$$X = \frac{\Delta H_m}{\omega_{PEEK}} \cdot \frac{1}{\Delta H_{th}} \cdot 100\ [\%] \quad (1)$$

where X is the degree of crystallinity [%], ΔH_m is the melting enthalpy [$J \cdot g^{-1}$], ω_{PEEK} is the mass fraction of PEEK of the composite and ΔH_{th} is the theoretical melting enthalpy of 100% crystalline phase of PEEK which is 130 $J \cdot g^{-1}$ [39].

2.3. Welding Processes

2.3.1. Ultrasonic Welding

One-mm thick unidirectional plates of CF/PEEK and 1mm thick PEI plates were used. Single-lap welded joints with the two different types of welding stacks sketched in Figure 2 were considered in this study. The first configuration was a direct weld between PEI and CF/PEEK. The second configuration consisted of two CF/PEEK plates with a PEI film as an energy director. It was a 250 μm thick flat energy director, which was placed at the welding interface before the welding process. It concentrated heat generation at the welding interface through combined surface friction, i.e., friction between the energy director and composite plates moving relative to each other, and viscoelastic friction, i.e., friction among adjacent polymer molecules in the energy director when subjected to cyclic deformation. This process provided welded joints over an entire overlapped surface. The specimens were welded in a single lap configuration (according to ASTM D1002) with a 10 mm long overlap.

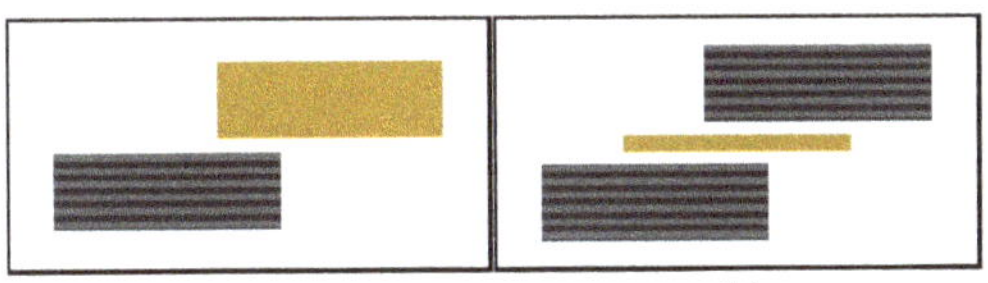

(a) (b)

Figure 2. Single-lap welded joints for USW (**a**) welding of PEI on CF/PEEK and (**b**) welding of CF/PEEK on CF/PEEK with 250 μm thick PEI as energy director (yellow bars are for PEI, grey bars for CF/PEEK).

Electrical Motion 20 ultrasonic welder from Rinco Ultrasonics (Romanshorn, Switzerland) was used for the experiments. An ultrasonic welding equipment was used to transmit mechanical vibration at a high frequency to the joint interface along with a static compressive force. The working frequency is 20 kHz and the maximum load of 3000 N. Amplifications provided by the booster and the titanium sonotrode were, respectively, 2 and 4. A clamping tool was designed to provide vertical movement without parasite bending to ensure pure friction during the welding test. Three welding parameters were considered in this study: the welding load of 500 N, the vibration amplitude of 32 μm, and the welding time of 1000 ms. The welding time is associated with the time when the ultrasound is operating. The welding load continued to be applied during the selected 2500 ms cooling phase.

2.3.2. Induction Welding

1 mm thick plates with cross stacking of CF/PEEK and 1 mm thick PEI plates were used. Single-lap welded joints with the two different types of welding stacks sketched in Figure 3 were considered in this study.

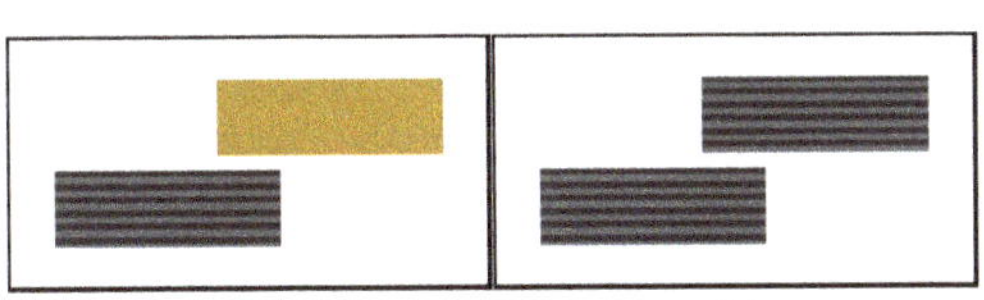

(a) (b)

Figure 3. Single-lap welded joints for IW (**a**) welding of PEI on CF/PEEK and (**b**) welding of CF/PEEK on CF/PEEK (yellow bar is PEI, grey bars for CF/PEEK).

In the case of PEI on CF/PEEK welds, the inductor was placed 3 mm above the area to be welded. A power of 55–70% and movement of the tool along the weld line at a speed of 2 mm·s^{-1} were used. During the welding of two CF/PEEK plates, the inductor was placed 2 mm above the surface of the welded area. The assembly cycle consisted of gradually increasing the power of the inductor to reach T_m + 20 °C at the start of the weld zone. For this purpose, a power of 80% and movement of the tool along the weld line at a speed of 3 mm·s^{-1} were used.

The welding tool consists of a CEIA 17 kW head (Paris, France) fitted with a 30 mm diameter pancake type inductor located at 2 mm or 3 mm. The inductor was preceded and followed by compaction rollers mounted on pneumatic cylinders. Two orientable Vortex fans (Terrebonne, QC, Canada) allow surface cooling of the weld area or its perimeter The welding tool is mounted on an ABB IRB6620 6-axis robot (Zurich, Switzerland). The specimens were positioned on top of each other with a 20 mm long overlap and held together with Kapton adhesive. The orientation of the folds at the weld was the same. N-type thermocouples were placed at the start and end of the weld to monitor the temperature of the interface. An induction welding was conducted without a susceptor at the interface, with carbon fibres replacing a metallic insert. An induction welding without a susceptor requires 45° lay-ups on the surface and depends on the electrical and thermal conductivity of the carbon fibre specimens. Induction welding was performed at IRT Saint Exupéry, Toulouse, France.

2.3.3. Transmission Laser Welding

Unidirectional plates of CF/PEEK 1 mm thick, PEI 1, 2, and 4 mm thick plates were used. Single-lap welded joints are sketched in Figure 4. The specimens were positioned on top of each other with a 20 mm long overlap and were held in position employing a pressure of 3.8 MPa applied through a glass plate to the overlap zone. Through transmission laser welding depends on the optical properties of the upper material.

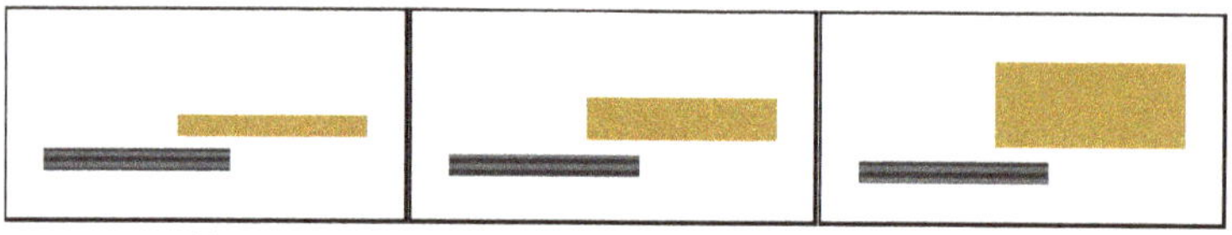

Figure 4. Single-lap welded joints for LW (**a**) welding of PEI 1 mm thick on CF/PEEK, (**b**) welding of PEI 2 mm thick on CF/PEEK and (**c**) welding of PEI 4 mm thick on CF/PEEK (yellow bars are PEI, grey bars for CF/PEEK).

PEI on CF/PEEK assemblies was made with a welding machine ES Weld 2000 by ES LASER (Bordeaux, France) equipped with an infrared laser beam at a wavelength of 970 nm with a maximum power of 225 W. The ovoid shape of the beam, about 5 mm wide, was obtained by a "top hat" lens allowing a uniform distribution of the laser radiation on the area to be welded. Two scans of the laser beam were arranged at an interval of 5 mm from each other. The laser speed was set up at 17 mm·s^{-1} and was carried out over the entire width of each assembly.

During these tests, the laser power was set between 20% (~34 W) and 35% (~69 W), and the laser scanning speed was between 17 mm·s^{-1} and 33 mm·s^{-1}. Visually, the welds were homogeneous for a scanning speed of 17 mm·s^{-1} and a power of 25% (~45 W), 30% (~52 W), and 35% (~69 W) for respectively the 1, 2, and 4 mm thick PEI samples. Then, the power used for each PEI thickness was adjusted more precisely: 45 W for 1 mm thick PEI, 50 W for 2 mm, and 66 W for 4 mm thick.

2.4. *Characterization of Weld Assemblies*

2.4.1. Mechanical Properties

The weld resistance was measured by the single lap shear test, where the maximum force was obtained after applying a tensile load to the specimens. The welds and test were performed following the ASTM D1002 standard, with 100×25 mm^2 samples corresponding to long and wide, respectively, and a minimal overlap area of 10×25 mm^2, on an Instron universal testing machine at 2800 N·min^{-1} cross-head speed. The lap shear strength was then calculated by Equation (2).

$$LSS = \frac{F_{max}}{A} \, [MPa] \tag{2}$$

In the lap shear test, the strength of a bonded joint (shear stress) depended on the load to the point of fracture (F_{max}) and overlapping area (A).

2.4.2. Optical Observation

The internal quality of the joints was inspected through a high-resolution microscope Keyence VHX-6000S (Bois-Colombes, France) in a reflection and transmission mode. Three-dimensional imaging was used for mapping each fractured interface after the mechanical test. Additional images were obtained from a camera Huawei P40 (Shenzhen, China) for a larger visual field. Image analysis of the welded surfaces was carried out using ImageJ v1.51k software (Wayn Rasband National Institutes of Health, Bethesda, MA, USA).

3. Results and Discussion

3.1. *Material Properties*

Both PEI and PEEK are sensitive to water intake, which affects the quality of the weld. If water is present, it evaporates upon heating, inducing material swelling and defects. Potente et al. [38] welded undried and pre-dried PEEK, they noticed bubbles in the undried PEEK welds only. Besides, heat damages were reported [33] on amorphous PEEK at a power of 20 W with a focal plane speed of 4 mm·s^{-1}. Although the authors agree about the role of water on defects, the effect of process parameters, such as power and sample speed, are modified with water content. The combined effect of water and fast heating ramp looks to worsen the apparition of defects. Welding processes are very energy-intensive and heating ramps are very high, possibly higher than 1000 °C·s^{-1}. Moreover, in transmission laser welding, heating is precisely located in a line or small points. Therefore, the presence of water in the materials causes unsymmetrical swelling of the substrates or locally, the creation of trapped bubbles due to the vaporization of water during welding.

Figure 5 presents the mass evolution for the sorption and desorption (drying) of PEI plates. Figure 5a shows the stabilization of the water content after 312 h of immersion. The corresponding saturation uptake was 1.30%wt. Then, the drying step in Figure 5b verifies the time required to remove water from PEI. During the first hours of drying, the weight decreased to −1.25%. The weight reduction was slower during the next hours, with an equilibrium after 8 h at a total mass loss of 1.40%. This weight loss was higher than the initial water uptake of 1.30%wt. This is due to the uncompleted drying of the initial state. As explained in Section 2.2.1, the first step was drying at 150 °C for 3 h: this duration was not enough to reach a fully dried state. Finally, our water equilibrium concentration is similar to Merdas' result of 1.39%wt. at 20 °C to 1.50%wt. at 100 °C [40].

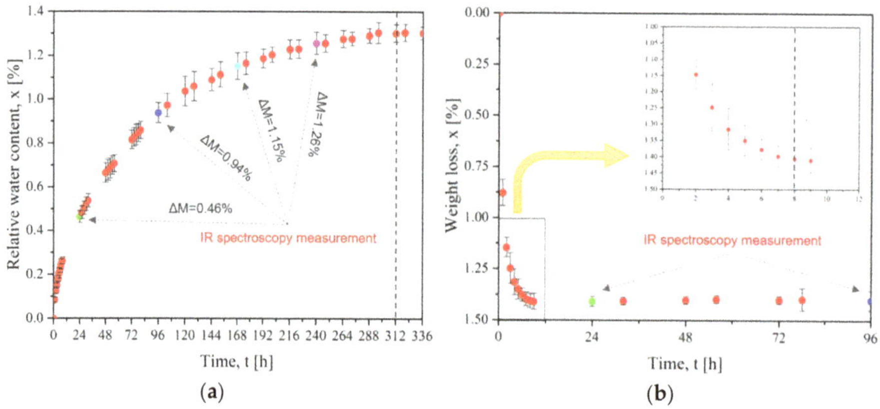

Figure 5. For PEI (**a**) Evolution of water diffusion at 25 °C and (**b**) Evolution of mass loss with drying time at 150 °C (dots of different colors represent measurement by FTIR spectroscopy).

Similarly, the evolution of the weight loss for CF/PEEK plates is presented in Figure 6. Figure 6a shows the diffusion of water into the CF/PEEK plates to achieve water saturation after 600 h (25 days) for a maximum mass gain of 1.17%wt. Figure 6b presents the drying curve of CF/PEEK: a rapid mass loss took place during the first hour and then the mass loss was slower. The equilibrium was reached after 24 h for a total mass loss of 1.24%. Grayson and Wolf [41] measured a water intake of 0.44%wt. at 35 °C to 0.55%wt. at 95 °C.

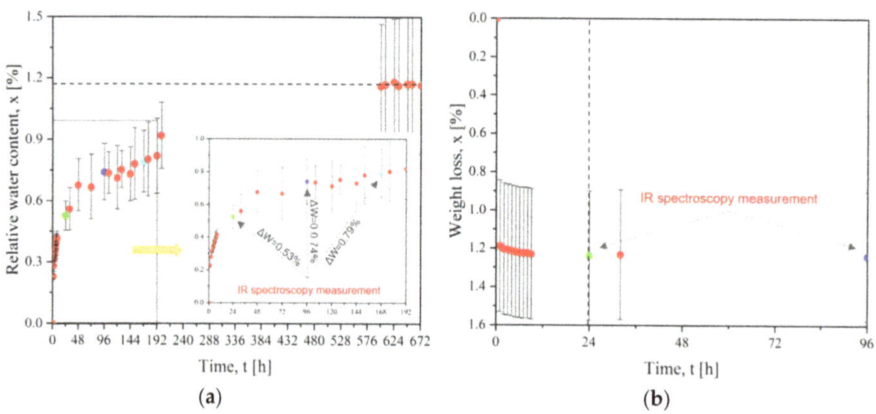

Figure 6. For HM63/PEEK (**a**) Evolution of water diffusion with time at 25 °C and (**b**) Evolution of mass loss with drying time at 150 °C (dots of different colors represent measurement by FTIR spectroscopy).

The chemical analysis of dried and wet samples was performed using Fourier-transform infrared spectroscopy which is based on the absorption of infrared radiation by bonds. The position and shape of absorption peaks in a spectrum indicate the nature of chemical groups characterizing the material. The spectra are provided as Supplementary Materials, Figure S1. For PEI, no structural change or degradation was observed after 24 h of sorption or drying. A slight change appeared after 96 h of drying when the -CH_2 groups were more visible in the 2700 cm^{-1} and 2750 cm^{-1} bands. Correa et al. [42] followed the evolution of stretching bands of -OH groups with sorption time to determine the dynamics of the transport of water molecules within PEI. He revealed different types of hydrogen bonding. He assigned the sharp peaks at 3655–3562 cm^{-1} to isolated water molecules interacting via H-bonding with the PEI backbone. A second doublet at 3611–3486 cm^{-1} was associated with water molecules self-interacting with the first shell species through a single H-bonding (self-associated or second-shell water molecules).

In the case of CF/PEEK, after 24 h of sorption, a strong stretch of -OH groups appeared at 3390 cm^{-1}. As sorption progresses, the spectrum was substantially similar with the intensity of the absorption peak increasing with water uptake. After 24 h of drying, the -OH groups disappeared, which means that water was reversibly removed from the material. Thus, for both materials, the heating time or water exposure was not sufficient to cause high levels of structural modification or chain breakage.

Based on these results, the drying conditions were defined as at least 8 h for PEI and 24 h for CF/PEEK.

The DSC and TGA thermograms of PEI and CF/PEEK are presented in Figures 7 and 8, respectively. The thermal transitions such as glass transition (T_g), melting (T_m) and crystallization temperature (T_x) were obtained from DSC. TGA thermograms give weight loss when the temperature increases. As a convention, the degradation temperature (T_d) was considered when the polymer lost 5% of its mass.

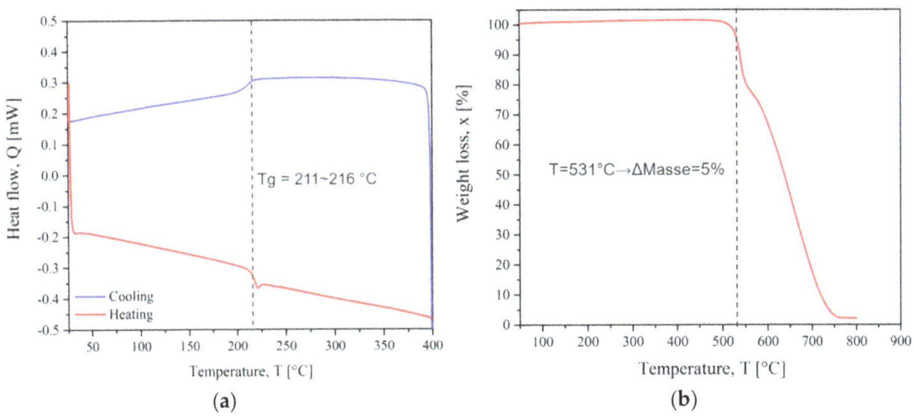

Figure 7. PEI thermograms (**a**) DSC and (**b**) TGA.

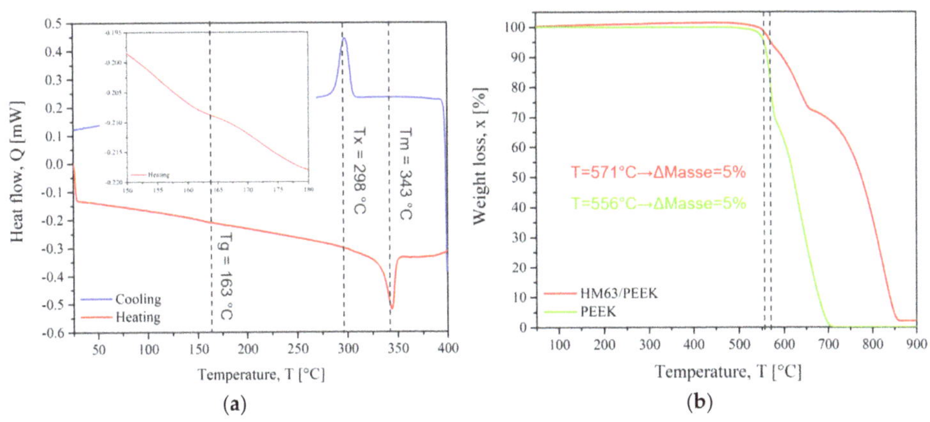

Figure 8. CF/PEEK thermograms (**a**) DSC and (**b**) TGA.

The T_g of PEI was observed between 211 °C and 216 °C. This reflects the change from the glassy state to the rubbery state in which the polymer chains gain mobility. The degradation took place in two main steps up to a complete weight loss. The degradation temperature was 531 °C. This temperature should not be exceeded to avoid damaging the chemical structure. Augh et al. [43] related no changes in molecular weight, glass transition temperature, or mechanical properties of PEI during an induction process. Because the time scale in the induction heating process was short, almost no degradation was observed Amancio Filho et al. [44] indicated that the primary thermal degradation mechanism of PEI is chain scission, with a very small decrease in average molecular weight. This minor decrease in molecular weight could be considered irrelevant for the mechanical performance of the joints because it is either above or within the molecular weight range where the strength of the PEI is independent of the variations in this property. So, we assume that the PEI macromolecular structure was not significantly modified by the welding cycles.

For CF/PEEK, the T_g of PEEK was estimated at 163 °C. The glass transition of pure PEEK is usually near 150 °C [45]. A slight modification of the macromolecules could stem from the interaction of PEEK macromolecules with carbon fibre surfaces [46]. The maximum melting peak was 343 °C. After melting, CF/PEEK crystallized on cooling with a maximum crystallization peak at 298 °C. The melting enthalpy was measured at 15 $J \cdot g^{-1}$ on heating. Considering that the volume fraction of carbon fibre is 61% of the composite, it was possible to calculate the degree of crystallinity with Equation (1). The degree of it for PEEK was 30% which is lower than the highest values reported in the literature for pure PEEK, up to 40% [45]. However, carbon fibres, apart from crystallization initiation and nucleation that decrease the process duration, do not modify it in general [47].

In Figure 8b, the degradation temperature for CF/PEEK is 571 °C, while it is faster for pure PEEK and starts at 556 °C. The degradation of CF/PEEK composites occurred in three steps. When reaching 800 °C, the weight loss was only 55% due to the presence of carbon fibres whose thermal stability is much higher than those of polymers. The thermal degradation of PEEK and CF/PEEK composites was well described in Gaitanelis' work [48].

For welding, the interface temperature must reach the temperature at which molecular mobility allows the interdiffusion of the polymeric chains. For amorphous thermoplastics, the welding temperature is after the glass transition and most often it is necessary to reach T_g + 140 °C [49] for the polymer to flow sufficiently. For semi-crystalline thermoplastics, welding takes place at the melting temperature or a few degrees above T_m + 70 °C [49], while preventing material degradation.

Based on these results, the theoretical welding range could be defined between 350 °C (T_g of PEI + 140 °C) and the degradation temperature of PEI: 350 °C < T < 530 °C for PEI on CF/PEEK assemblies. The lowest temperature limit, 350 °C corresponds also to the T_m of CF/PEEK.

The thermal conductivity of PEI was 0.20 $W \cdot (m \cdot K)^{-1}$, and the specific heat was 1068 $J \cdot (kg \cdot K)^{-1}$. Comparatively, the through-plane thermal conductivity obtained for CF/PEEK was 15.95 $W \cdot (m \cdot K)^{-1}$ and up to 55.00 $W \cdot (m \cdot K)^{-1}$ in plane one. Thus, PEI acts as a thermal insulator in welding, keeping heat inside the material. Oppositely, in CF/PEEK, the fibres contribute to fast heat dissipation inside the whole elements to be welded. The thermal conductivity of CF/PEEK is higher than those of PEI, so the heat is dissipated faster in composites [50]. Regarding induction welding, the fibres play the role of a conductive element, allowing for the generation of Eddy currents inside the composite instead of a metallic susceptor. Rudolf et al. [51] reported that continuous carbon fibres may act similar to closed loops within the composite structure.

The dynamic mechanical analysis gives information on the macromolecular mobility and the temperature processing window. The welding configurations were either CF/PEEK on CF/PEEK, PEI on CF/PEEK and CF/PEEK on CF/PEEK with a 250 μm thick interfacial PEI film.

Polymers are immiscible in most cases, but PEEK and PEI are fully miscible according to Hsaio [36]. This makes PEI a relevant option to assemble PEEK composites.

In Figure 9a, the storage modulus (E') of PEEK shows three distinct regions: a high modulus glassy region where the segmental mobility is restricted, a transition zone where a substantial decrease of E' with the increase in temperature, and a rubbery region (the flow region) where a drastic decay is seen.

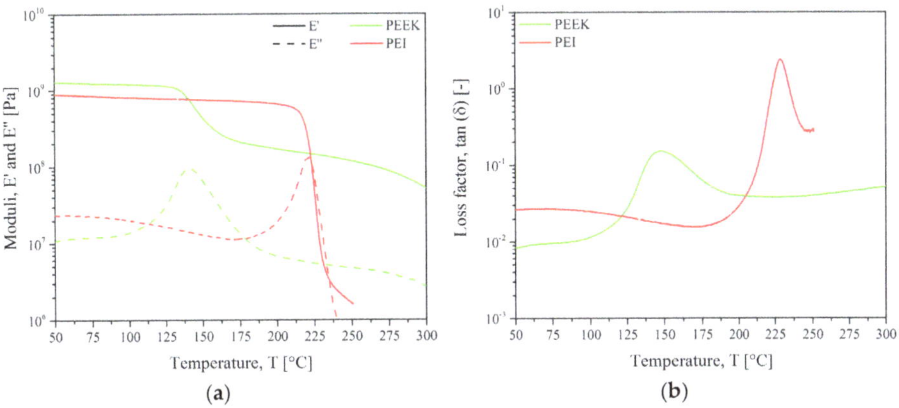

Figure 9. Dynamic mechanical analysis for pure PEEK and PEI (**a**) storage and loss moduli and (**b**) loss factor.

The E′ curve of PEI shows the typical behavior of an amorphous polymer, which migrates from an energy-elastic to an entropy-elastic state after reaching the glass transition temperature. The amorphous plateau was not visible: above 250 °C, the PEI was so soft that measuring is not possible in this configuration. Instead, a plate configuration would be suitable. For both materials, the loss modulus (E″) curves demonstrate broad and asymmetrical peaks related to the energy dissipated through inter macromolecular frictional motion.

Figure 9b indicates the mechanical response of the glass transition (T_α) when tan(δ) reaches a maximum. The peak of PEEK was broader and less symmetrical than those of PEI. The peak shape is correlated with the polydispersity and the various relaxation times characterizing macromolecular mobility.

The storage modulus of PEI dropped right after its glass transition at 217 °C, giving the macromolecules a high level of mobility while PEEK chains kept their rigidity. So, PEI would facilitate the motion and the diffusion of PEEK chains across the interface. Thus, less energy would be required to soften the PEEK matrix. Based on these results, PEI could be assembled with PEEK from 250 °C when the PEI macromolecules gained enough mobility to diffuse within PEEK.

For PEEK/PEEK welding, previous studies demonstrated that it is necessary to attain its melting temperature close to 360 °C for effective interfacial bonding [35,52,53].

The transmission coefficient of PEI and CF/PEEK was measured to check their weldability in laser through-transmission configuration in the wavelength range of 400 nm to 1000 nm. For a fixed temperature, the transmittance is influenced by the chemical structure, surface roughness, angle of incidence of the light, and sample thickness. The measurements were carried out at 20 °C for 1, 2, and 4 mm PEI plates as well as for CF/PEEK plates in Figure 10a. The evolution of the transmittance with temperature for PEI was followed in the range of 20–140 °C in Figure 10b.

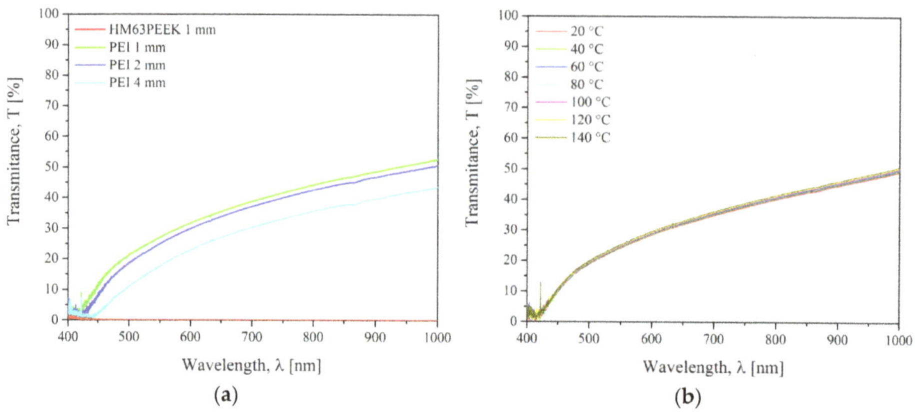

Figure 10. Transmission coefficient of (**a**) PEI and HM63/PEEK at 20 °C and (**b**) PEI for different temperatures.

The transmission of PEI was negligible close to 400 nm, then it increased to 57% for 1 mm thick at a maximum of 1000 nm. At the wavelength of interest at 970 nm, the transmittance was 52%, 50%, and 43% at 20 °C for 1, 2, and 4 mm thick samples, respectively. As expected from the Beer-Lambert law presented in Supplementary Materials, Figure S2, the transmission factor increases when the thickness of the sample decreases. It is well-known that the transparency of amorphous polymers increases the scatter of the laser light, which in turn increases the effective beam diameter [54]. For the CF/PEEK, the transmission was 0% over an entire wavelength range. This is due to the high absorption ability of the carbon fibres. Chabert et al. [16] reported that below 23% of transmission, the power rate of the laser reaching the interface was too low. The energy density at the interface was not high enough to allow polyamide to reach its melting temperature. In Figure 10b, it is observed that the thermal dependence of the transmittance of PEI was negligible in the range of 20 °C to 140 °C. Since there is no thermal transition, it is expected the same transmittance for PEI upon heating up to T_g = 217 °C. For a through transmission laser welding, the upper substrate must be semi-transparent to the laser wavelength, whereas the lower substrate must be opaque. Thus, the beam energy needs to be completely absorbed by the lower substrate which will heat up, and by conduction, will cause the temperature of the upper substrate to rise. According to the transmission factors, PEI plates could be assembled as an upper element by a through transmission laser welding whatever their thickness up to 4 mm. CF/PEEK is suitable as a lower element.

3.2. *Interfacial Strength of Assemblies*

3.2.1. Monitoring of the Interfacial Temperature

To ensure a satisfactory interfacial adhesion, it is necessary to reach a temperature above the melting temperature of PEEK or the glass transition of PEI. Controlling temperature at the interface during welding, therefore, would ensure the quality of the welds. Since temperature is a key parameter for optimizing welding processes, the temperature at the interface during heating and cooling was monitored.

The temperature at the interface was measured during welding for the three processes. For ultrasonic welding, K-type thermocouples were inserted inside the PEI energy director film. The results were reported in Bonmatin's work [27].

For induction welding, N-type thermocouples were placed at the start and end of the weld, the results are shown in Figure 11.

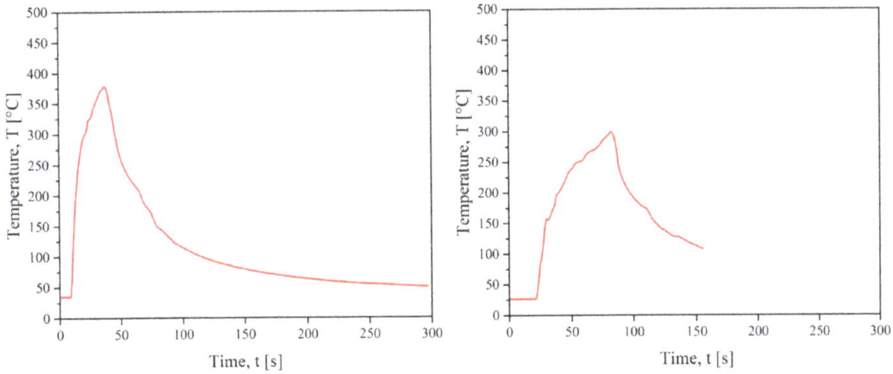

Figure 11. Temperature monitoring during IW for (**a**) CF/PEEK on CF/PEEK and (**b**) PEI 1 mm on CF/PEEK.

The maximum temperature reached for welding CF/PEEK on CF/PEEK was higher than the melting temperature of PEEK. It took about 30 s to achieve the maximum temperature at 375 °C (Figure 11a). Besides, a lower temperature of 300 °C was sufficient for welding PEI on CF/PEEK (Figure 11b), which was higher than the glass transition of PEI at 216 °C. This result is consistent with our hypothesis explaining that PEI chains facilitate the motion and the diffusion of PEEK across the interface. Thus, less energy is required to soften the interface.

During transmission laser welding, three K-type thermocouples were placed in the PEI (upper element) close to the interface, at the start, in the middle, and at the end of the weld.

Even though temperature monitoring is recognized as essential to improving the quality of welds, in-situ measurements are rare in the literature. The obstacles come from the closed interface and the displacement of the contact surface in the case of ultrasonic welding. The main approach consists of embedding a thermocouple inside an interfacial polymer layer [27] or inside a part to be welded as close as possible to the contact surface [55]. Another option is to measure the temperature distribution by infrared thermography with the camera field either perpendicular [17] or parallel to the welded interface [56–59]. Then, fiber Bragg grating sensors are constructed in a short segment of optical fibre that reflects particular wavelengths of light and transmits all others. They are used to monitor physical parameters such as temperature even in inaccessible, unconventional environments. One such application is monitoring the temperature of a substrate from the back surface such as any conventional thermocouple sensor [60,61]. In the latter case, numerical models were applied to calculate the interfacial temperature taking into account the thermal properties of materials [62].

For welding processes hereby considered, temperature ramps measured with thermocouples are gathered in Table 1.

Table 1. Heating and cooling temperature ramps for welding processes.

In $K{\cdot}s^{-1}$	Ultrasonic Welding	Induction Welding	Laser Welding
Heating ramp	1000	10	200
Cooling ramp	500	5	100

Thus, the fastest temperature increase was for ultrasonic welding. Whatever the process, we assume the maximum temperature at the interface reached the thermal transitions of PEI and PEEK as defined by DSC without exceeding the degradation temperature determined by TGA. This assumption was based on an interfacial resistance obtained from mechanical tests presented below.

We underline that such degradation temperature was measured in slow ramps (5 K·min^{-1}). When PEEK and CF/PEEK are exposed to rapid heating rates, they withstand much higher temperatures up to 600 °C without evidence of degradation, as attested by Gaitanelis [48] and Bonmatin's work [27].

3.2.2. Mechanical Resistance of Welded Parts

The resistance of welded specimens was probed by single lap shear tests. The strength (LSS) values are reported in Figure 12. Three configurations were explored as reported in Figures 2–4 for an ultrasonic, induction and laser welding respectively. For ultrasonic and laser processes, the assemblies were obtained by overlapping two 25 mm width specimens For induction welding, two 50 mm width specimens were overlapped to fit our induction welding set-up. Then, they were cut to provide the 25 mm width assemblies. Some of them were separated during cutting, which reflected the low resistance of the welds: these specimens were discarded. The nomenclature is either material X/PEI/material Z when a PEI film was placed at the interface or material X/material Z without an interfacial film. For example, HM63PEEK/PEI/HM63PEEK refers to a joint between two elements of CF/PEEK with a PEI energy director. At least 6 assemblies welded with identical parameters were tested. Each point of the graphs corresponds to an assembly, whereas the bars are the average value of these points, with the standard deviation.

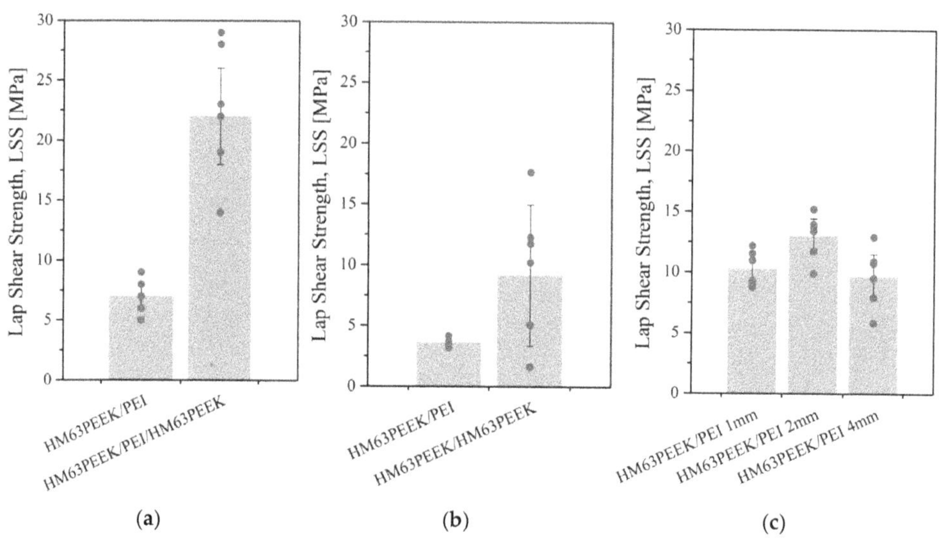

Figure 12. Lap shear strength of test specimens after (**a**) ultrasonic, (**b**) induction and (**c**) transmission laser welding. LSS is calculated from the effective welded surface.

LSS was calculated from Equation (2) by dividing the force to separate the welded specimens by the welded area. For ultrasonic welding, the welded areas were considered to be the fully overlapped surface: 25 × 10 = 250 mm^2. When observing the fractured surface, the overlapped surface looked totally melted, which strengthened our choice. Oppositely, the welded areas obtained by induction and laser transmission were theoretically 25 × 20 = 500 mm^2. However, because the melted zone did not correspond to the entire surface, each surface was measured precisely to ensure a reliable LSS value. The welded areas obtained by an induction and laser transmission were calculated using the software ImageJ on PEI surfaces. The areas could not be measured on composite surfaces because the melted zones were not clearly visible on black surfaces. In the case of induction, the size of the welded areas was from 160 mm^2 to 290 mm^2

for PEI on CF/PEEK assemblies and from 460 mm^2 to 520 mm^2 for CF/PEEK on CF/PEEK assemblies, meaning that 32% to 100% of the overlapped surface was effectively welded. Laser welding provides a very repetitive bonded area. The geometries were two parallel lines apart to create one large welded zone. The width of one line was estimated between 4 mm and 4.5 mm, whereas the distance of laser passage between them was set up at 6 mm. The variability of the welded surfaces was very low, with an average value of 210 $\pm$ 5 mm^2 corresponding to 42% of the overlapped surface, regardless of the thickness of the upper substrates. This highlights the reproducibility of the welds produced by a through transmission laser welding.

In Figure 12a,b, the values obtained for PEI on CF/PEEK assemblies correspond to a 1 mm thick PEI substrate. In addition, a PEI film was used as an energy director in ultrasonic welding for CF/PEEK on CF/PEEK welds. In the case of laser welding, 1, 2, and 4 mm thick PEI substrates were assembled on 1 mm thick CF/PEEK.

For some 1 and 2 mm thick PEI on CF/PEEK assemblies, the fracture took place in the PEI specimen instead of at a welded interface. Therefore, the LSS results were not the resistances of the weld but those of the PEI substrate. These values indicate that the weld was stronger than the maximum stress acceptable by the PEI plates. This case is discussed below. The highest LSS mean value was noticed for ultrasonic welding for welds between two CF/PEEK composites at 22 MPa. The best connection between the PEI and CF/PEEK was obtained for transmission laser welding with 2 mm thick PEI at a mean value of 13 MPa. With the same configuration, the LSS obtained for ultrasonic welding was 7 MPa. For both configurations, the lowest values of LSS were obtained for induction welding with an average of 10 MPa and 4 MPa for CF/PEEK on CF/PEEK and PEI on CF/PEEK, respectively.

The narrowest standard deviation was obtained for laser welding. The highest standard deviation was for induction welding when joining CF/PEEK on CF/PEEK and for an ultrasonic welding when joining CF/PEEK on CF/PEEK with a PEI energy director. However, the highest LSS values were obtained with the latter process/material couple with about 30 MPa. This result could stem from the difficulty to prevent the PEI interfacial layer from flowing. Indeed, a reasonable flow is beneficial to the mechanical resistance, whereas an excessive flow of PEI induces a fibre displacement and deformation of the specimens, resulting in a lower strength.

To sum up, ultrasonic welding seems to be well-suited for assembling composite on composite whereas transmission laser welding is appropriate for PEI/composite welds.

The analysis of fracture modes is presented below based on photographs of assemblies after the mechanical tests. The images of fracture surfaces for an ultrasonic, induction, and laser welding are presented in Figure 13, Figure 15 and Figure 16, respectively.

- Ultrasonic welding

For ultrasonic welding, in Figure 13a, a slight fibre distortion at the edges of the CF/PEEK on CF/PEEK welds (with a PEI film at the interface) was observed, the latter could be due to the welding load of 500 N applied up to the cooling. A significant flow of the PEI was observed with slight deformation of the fibres on the edges of the assemblies. A mixed cohesive/adhesive failure occurred at the interface and in the inner layers of the two substrates. In Figure 13b, the welding of PEI on CF/PEEK resulted in a break in the PEI substrate. This indicates that the resistance of the interface was greater than that of the PEI substrate, whose thickness was 1 mm in this case.

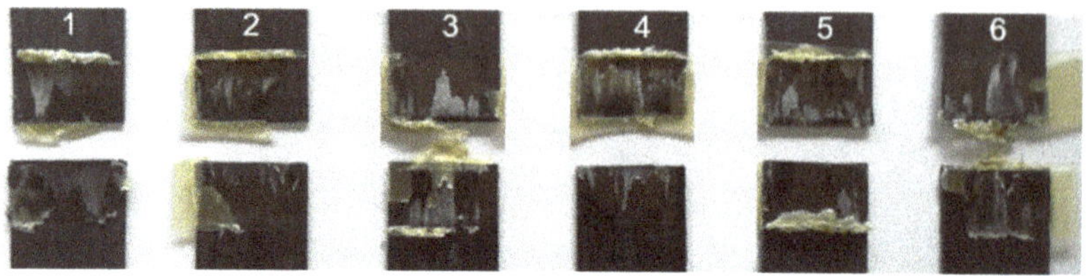

(**a**)

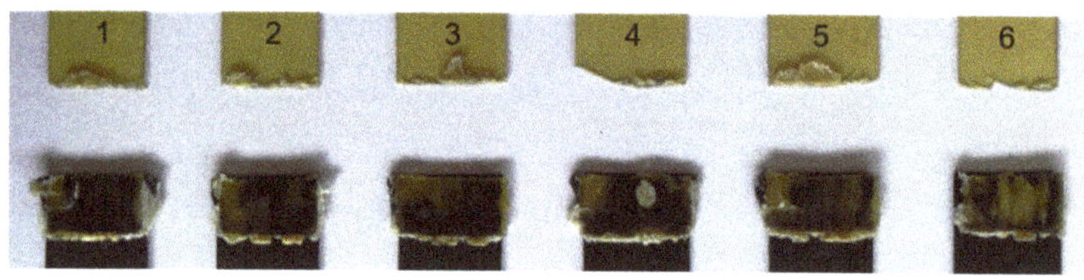

(**b**)

Figure 13. Fracture surfaces after LSS testing of ultrasonically welded specimens from 1 to 6 for (**a**) CF/PEEK on CF/PEEK with 250 μm thick PEI as energy director and (**b**) 2 mm thick PEI on CF/PEEK.

For further analysis, Figure 14 displays the interfacial fracture of a weld that exhibits both cohesive and adhesive failure within CF/PEEK plies. In Figure 14a, the optical micrograph from an entire welding area after a lap shear test is presented. An image of a specific location for two CF/PEEK plies is presented in Figure 14b. In that case, the fibres were pulled out, confirming that a cohesive fracture within one ply occured. Some fibres from the second ply were attached to the second composite specimen which is obvious from the height difference of 150 μm measured by optical microscopy. The layer on the right belongs to the first ply of the composite of 140 μm whereas the layer on the left belongs to the second one. A hypothesis is that cracks initiated in both CF/PEEK overlapped specimens of this assembly. The cracks crossed from one CF/PEEK layer to the next one through the PEI-rich interphase, giving rise to this rough surface.

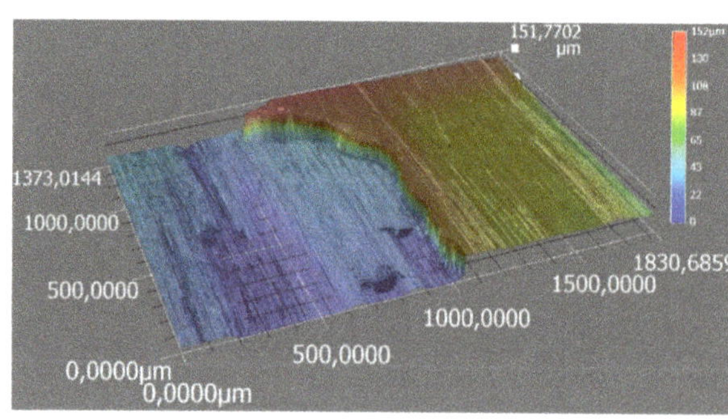

(**a**) (**b**)

Figure 14. Optical micrographs after LSS testing of ultrasonically welded of (**a**) interfacial fracture of two layers of CF/PEEK and (**b**) colour chart corresponds to a height difference.

With the PEI film at the interface, the fractured surfaces show PEI flowing along the edge of a transverse outer lap. Also, a fibre breakage in the middle of the lap was observed.

Further examination of the damaged areas on the fracture surface revealed the presence of dry carbon fibres on the edge of the transverse outer lap, which could be interpreted as a sign of thermal degradation. It is possible that the maximum temperature reached was locally too high.

- Induction welding

In the case of induction welding in Figure 15a, neither a fibre distortion nor polymer flow was noticed, ensuring accurate dimensional stability. For CF/PEEK on CF/PEEK, a mixed cohesive/adhesive failure occurred at the interface and the inner layers of both substrates. Because of black surfaces and slightly deformed samples, determining the welded area was not possible. For this reason, we considered the total overlapped area to calculate LSS, which underestimates the LSS value.

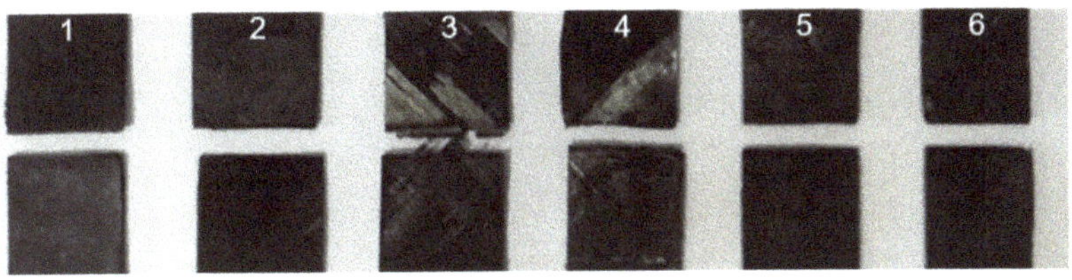

(**a**)

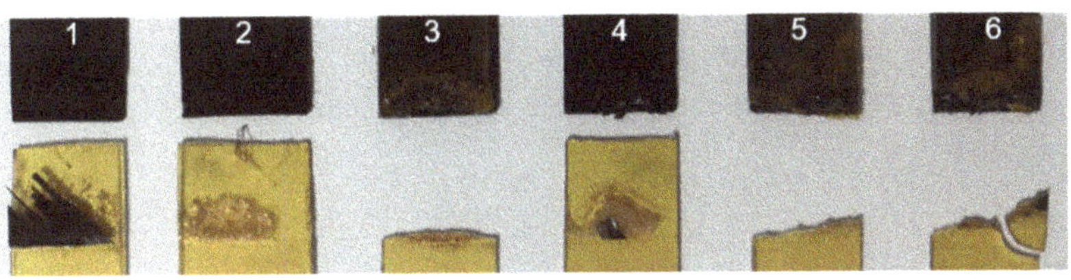

(**b**)

Figure 15. Fracture surface after LSS testing of induction welded specimens from 1 to 6 for (**a**) CF/PEEK on CF/PEEK and (**b**) 1 mm thick PEI on CF/PEEK.

Welds of PEI on CF/PEEK in Figure 15b allow observing a presence of bubbles or a grainy appearance on some specimens. This could come from a material flow during the application of pressure up to 1.5 MPa. Half of the 1 mm thick PEI assemblies broke within PEI, indicating that an interface resistance was higher than that of the PEI substrate. The others gave rise to an adhesive rupture, with the presence of a few fibres adhered to the CF/PEEK substrate.

- Through transmission laser welding

In Figure 16, the failure of the PEI substrate is observed for the assemblies with the 1 mm and 2 mm thick PEI. Again, this implies that the interface had an LSS higher than the resistance of the PEI substrate. In the case of the 4 mm thick PEI, a rupture of the interface was observed. Some fibres of the CF/PEEK remained on the PEI, indicating a cohesive failure.

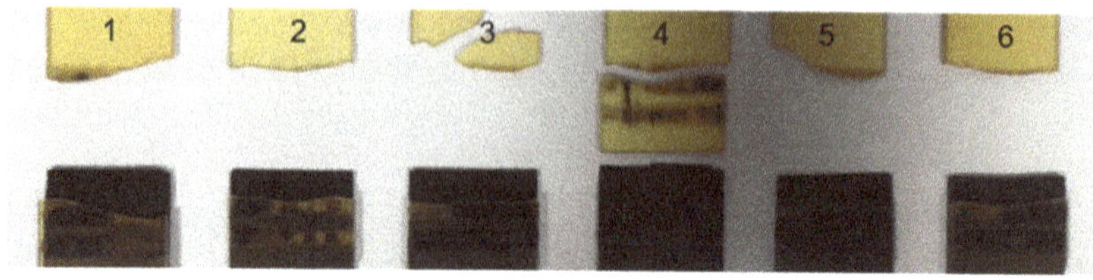

(a)

(b)

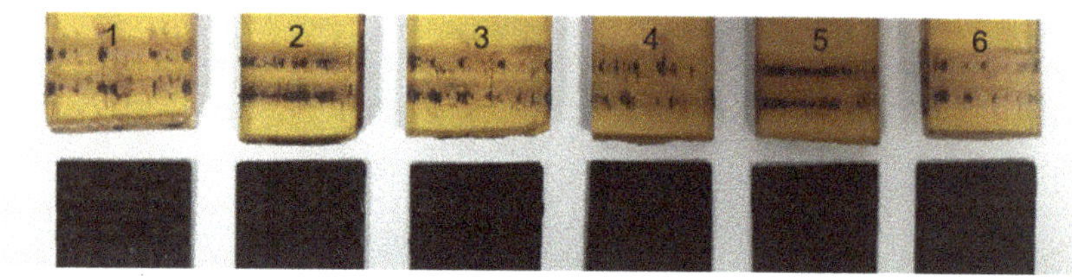

(c)

Figure 16. Fracture surface after LSS testing of transmission laser welded specimens from 1 to 6 for (**a**) 1 mm PEI, (**b**) 2 mm PEI and (**c**) 4 mm thick PEI on CF/PEEK.

Still, for laser welding, Figure 17 shows an optical micrograph from the entire welding area after a lap shear test. In both cases, it is possible to observe some fibres on the surface of the PEI substrate, proving a strong adhesion at the interface. The differences in failure and LSS results are significant: the average LSS was higher for 2 mm thick PEI than for 4 mm thick. Indeed, a weld of 4 mm PEI on CF/PEEK in Figure 17b looks degraded with a more uneven and darker surface compared to 2 mm PEI. Such thermal degradation may be caused by changes in energy density applied during the welding process. Indeed, for welds of 2 mm PEI on CF/PEEK and 4 mm PEI on CF/PEEK, the energy density of 14.7 $J{\cdot}mm^{-2}$ and 19.5 $J{\cdot}mm^{-2}$ were used, respectively. The energy density was increased to compensate for the decrease in transmission factor when the PEI is thicker, as seen in Section 3.1 in Figure 10. However, too high energy density may provoke the degradation of the material and result in lower mechanical properties.

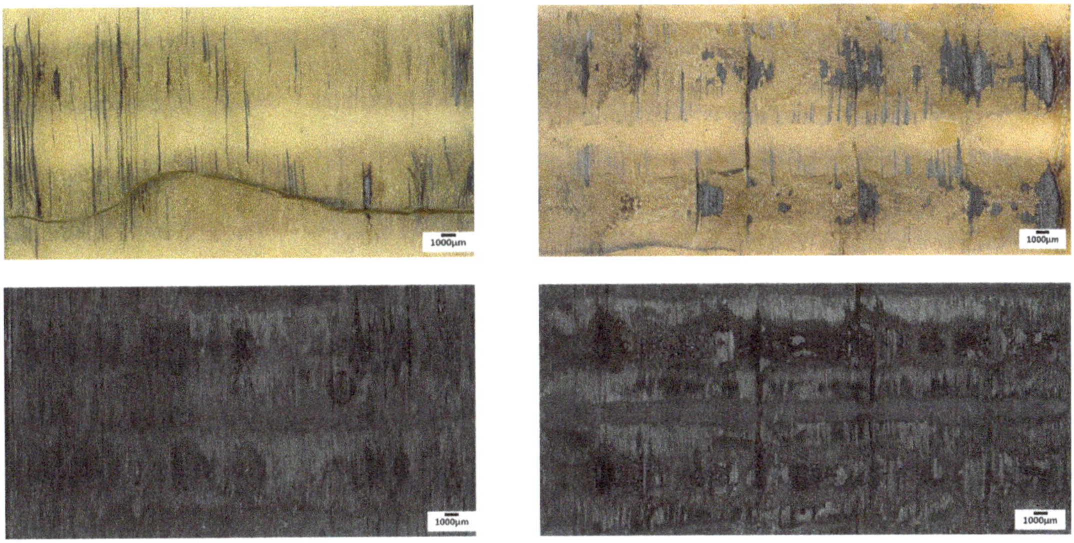

(a) (b)

Figure 17. Optical micrographs of interfacial fracture after LSS testing of transmission laser welded of (**a**) 2 mm PEI and (**b**) 4 mm thick PEI on CF/PEEK (on the top, transparent PEI, on the bottom, absorbent CF/PEEK).

- Summary

For PEI on CF/PEEK welds, the failure of PEI was observed for all welds whatever the welding process. Indeed, the PEI specimens were solicited in a tensile mode and when the load reached a tensile yield, the PEI specimen broke. This means that the welded joint was stronger than the PEI. The tensile properties of injection moulded PEI specimens were measured: the ultimate tensile strength was 99 MPa and the elastic modulus was 3300 MPa. The results obtained when the PEI substrate failed were determined by the tensile stress experienced by the PEI substrate. Considering the maximum force reached during the LSS test divided by the specimen section (width × thickness), an average value of 50 MPa was obtained; half of the ultimate tensile strength (UTS) of injection moulded PEI specimens. One could be noticed that the welded specimens were manufactured by compression moulding whereas UTS was obtained from injection moulded specimens. It is well-known that injection moulding orientates the macromolecules in the flow direction; the latter is also the testing direction. When pulled out, the intramolecular bonds—mainly C-C covalent bonds—are stretched. Oppositely, in compression moulding, the polymer conformation obeys the Gaussian random coil: there is no preferential orientation in their organization. During LSS testing, the intermolecular bonds—van der Waals and hydrogen bonds, whose dissociation energy is much lower—are also involved in the mechanical resistance.

Amanat et al. [33] reported different failure modes: interfacial, bulk substrate, near interfacial substrate, substrate, and interfacial. The mode of failure for joints was adapted from classifications used with adhesive-based joints [63,64]. In our case, the failure in the PEI substrate resulted in a tensile yield and break (Figure 13b, Figure 16a,b). When the thickness was 4 mm (Figure 16c), the PEI did not break, highlighting the weld resistance.

During ultrasonic welding, a PEI film inserted between the two CF/PEEK specimens strengthens the welds. Indeed, when assembling CF/PEEK on CF/PEEK, some fibre/fibre contact is possible in some areas, preventing a full close contact. Adding a PEI film ensures a polymer layer that is softened upon heating. However, it seems that the pressure was so high that this layer flowed beyond the specimen edges, as a consequence, dimensional accuracy was not kept.

On the contrary, dimensional accuracy was guaranteed for induction welding. However, the fracture surfaces between composites (Figure 17) showed damage in the CF/PEEK material along the transverse external overlap edge and also in the middle of the overlapped zone. A closer look into the damaged areas on the fracture surface revealed the presence of dried carbon fibres (without polymeric matrix) at the transverse external overlapped edge, which could be interpreted as a sign of thermal degradation. The middle of the overlap shows, however, fractures within a resin-rich interlayer.

4. Conclusions

The originality of this work is to compare ultrasonic, induction and transmission laser welding processes for the assembly of thermoplastic composites. The targeted application is a skin for sandwich panels for the space industry. The process parameters have still to be optimized, however, selected operation points give values of LSS high enough to fit some industrial specifications.

The materials were thoroughly characterized to highlight the specific properties involved in each welding process. Laser welding requires the upper element to be transparent to the laser wavelength. The transmission factor of PEI was 52%, 50%, and 43% at 20 °C for 1, 2, and 4 mm thick samples at the wavelength of interest at 970 nm. Induction welding required the material to be electrically and thermally conductive. The thermal conductivity of CF/PEEK was 55 $W{\cdot}(m{\cdot}K)^{-1}$, which makes it suitable to be welded by induction without a metallic susceptor. Ultrasonic welding is the most versatile process as it does not necessitate any material properties.

For CF/PEEK on CF/PEEK, the maximum LSS value of 28.6 MPa was reached for an ultrasonic welding, while an induction one brought 17.6 MPa. This difference is partially due to the addition of a 250 μm thick PEI film used as an interfacial energy director for ultrasonic welding. CF/PEEK on CF/PEEK joints resulted in mixed cohesive/adhesive failure at the interface and within the inner layers of both substrates.

For PEI on CF/PEEK assemblies, the three processes are suitable. The maximum LSS values were obtained for laser welding at 15.2 MPa, followed by ultrasonic welding at 8.5 MPa and induction welding at 4.2 MPa. These low values are explained by the fracture of PEI specimens when the PEI thickness was below 4 mm. The LSS values reported are not associated with the interfacial resistance but with the tensile strength of the PEI itself. The highest LSS for welds with PEI were obtained by laser welding for PEI of 2 mm thickness. It is worth highlighting the effect of the low thermal conductivity of PEI compared to CF/PEEK layers. The heat is dissipated faster inside the composite layers due to carbon fibres, which require reaching higher temperatures to bring enough heat to the interface in the case of CF/PEEK on CF/PEEK assemblies.

In all welding processes, controlling the interfacial temperature is not easily controlled. The temperature must stay above the glass transition of PEI or melting temperature of PEEK while preventing degradation and delamination of composite plies. Further research efforts are necessary to develop reliable in-situ temperature measurements to monitor an interfacial temperature, as a step towards dimensional accuracy and weld resistance. Further works will be oriented to gain insights into the effect of process parameters on temperature distribution in the contact area.

Finally, the three processes are promising for assembling carbon fibre/PEEK composites. The ongoing study aims to optimize the process parameters to increase the strength of the welds, before transferring the processes for assembling parts for space applications such as satellites.

Supplementary Materials: The following are available online at https://www.mdpi.com/article/10.3390/ma15186365/s1, Figure S1: FTIR spectra, Figure S2: Beer-Lambert law graph.

Author Contributions: Conceptualization, A.K., C.G., E.L. and F.C.; methodology, A.K., M.B., C.G. and F.C.; validation, C.G., E.L. and F.C.; formal analysis, A.K., C.G. and F.C.; investigation, A.K. and M.B.; writing—original draft preparation, A.K.; writing—review and editing, A.K., C.G., E.L. and F.C.; supervision, C.G., E.L. and F.C.; funding acquisition, C.G., E.L. and F.C. All authors have read and agreed to the published version of the manuscript.

Funding: This research was funded by CNES (French National Space Agency), grant number R&T R-S19/TG-002-082.

Data Availability Statement: The data that support the findings of this study are available from the corresponding author.

Acknowledgments: The authors would like to thank Pr Thierry Cutard and his team at ICA IMT Albi, France for the access to the ultrasonic welding machine and Pr Fabrice Schmidt and Rémi Gilblas, ICA IMT Albi, France for performing spectrophotometry measurements. We are grateful to Guillaume Morel, INP-ENIT, Tarbes, France, for helping to manufacture the plates and performing thermal conductivity measurements.

Conflicts of Interest: The authors declare that they have no known competing financial interest or personal relationships that could have appeared to influence the work reported in this paper.

References

1. Grimm, R.A. Welding processes for plastics. *Adv. Mater. Process.* **1995**, *147*, 27–30.
2. Lambiase, F.; Derazkola, H.A.; Simchi, A. Friction stir welding and friction spot stir welding processes of polymers—state of the art. *Materials* **2020**, *13*, 2291. [CrossRef] [PubMed]
3. Reis, J.P.; de Moura, M.; Samborski, S. Thermoplastic composites and their promising applications in joining and repair composites structures: A review. *Materials* **2020**, *13*, 5832. [CrossRef] [PubMed]
4. Kumar, R.; Singh, R.; Ahuja, I.P.S.; Penna, R.; Feo, L. Weldability of thermoplastic materials for friction stir welding—a state of art review and future applications. *Compos. B Eng.* **2018**, *137*, 1–15. [CrossRef]
5. Díaz, J.; Rubio, L. Developments to manufacture structural aeronautical parts in carbon fibre reinforced thermoplastic materials. *J. Mater. Process Technol.* **2003**, *143–144*, 342–346. [CrossRef]
6. Stavrov, D.; Bersee, H. Resistance welding of thermoplastic composites-an overview. *Compos. Part A Appl. Sci. Manuf.* **2005**, *36*, 39–54. [CrossRef]
7. Fernandez Villegas, I. Ultrasonic welding of thermoplastic composites. *Front. Mater.* **2019**, *6*, 291. [CrossRef]
8. Bhudolia, S.K.; Gohel, G.; Leong, K.F.; Islam, A. Advances in ultrasonic welding of thermoplastic composites: A review. *Materials* **2020**, *13*, 1284. [CrossRef]
9. Qiu, J.; Zhang, G.; Sakai, E.; Liu, W.; Zang, L. Thermal welding by the third phase between polymers: A review for ultrasonic weld technology developments. *Polymers* **2020**, *12*, 759. [CrossRef]
10. Tao, W.; Su, X.; Wang, H.; Zhang, Z.; Li, H.; Chen, J. Influence mechanism of welding time and energy director to the thermoplastic composite joints by ultrasonic welding. *J. Manuf. Process.* **2019**, *37*, 196–202. [CrossRef]
11. Ahmed, T.J.; Stavrov, D.; Bersee, H.E.N.; Beukers, A. Induction welding of thermoplastic composites—an overview. *Compos. Part A Appl. Sci. Manuf.* **2006**, *37*, 1638–1651. [CrossRef]
12. Dave, F.; Ali, M.M.; Sherlock, R.; Kandasami, A.; Tormey, D. Laser transmission welding of semi-crystalline polymers and their composites: A critical review. *Polymers* **2021**, *13*, 675. [CrossRef] [PubMed]
13. Ageorges, C.; Ye, L.; Hou, M. Advances in fusion bonding techniques for joining thermoplastic matrix composites: A review. *Compos. Part A Appl. Sci. Manuf.* **2001**, *32*, 839–857. [CrossRef]
14. Duley, W.W.; Mueller, R.E. CO_2 laser welding of polymers. *Polym. Eng. Sci.* **1992**, *32*, 582–585. [CrossRef]
15. Labeas, G.N.; Moraitis, G.A.; Katsiropoulos, C.V. Optimization of laser transmission welding process for thermoplastic composite parts using thermo-mechanical simulation. *J. Compos. Mater.* **2010**, *44*, 113–130. [CrossRef]
16. Chabert, F.; Garnier, C.; Sangleboeuf, J.; Akue Asseko, A.C.; Cosson, B. Transmission laser welding of polyamides: Effect of process parameter and material properties on the weld strength. *Procedia Manuf.* **2020**, *47*, 962–968. [CrossRef]
17. Villar Montoya, M.; Garnier, C.; Chabert, F.; Nassiet, V.; Samélor, D.; Diez, J.C.; Sotelo, A.; Madre, M.A. In-situ infrared thermography measurements to master transmission laser welding process parameters of PEKK. *Opt. Lasers Eng.* **2018**, *106*, 94–104. [CrossRef]
18. Jasiulek, P. Przykłady przemysłowego zastosowania łączenia tworzyw sztucznych. *Spajanie Metali i Tworzyw w Praktyce* **2006**, *4*, 21–24.
19. Deng, S.; Djukic, L.; Paton, R.; Ye, L. Thermoplastic—epoxy interactions and their potential applications in joining composite structures—A review. *Compos. Part A Appl. Sci. Manuf.* **2015**, *68*, 121–132. [CrossRef]
20. Rydzkowski, T.; Michalska Pożoga, I. Rozwój technik łączenia materiałów—spawanie tworzyw polimerowych. *Weld. Tech. Rev.* **2015**, *87*, 18–21. [CrossRef]

21. Fernandez Villegas, I. Strength development versus process data in ultrasonic welding of thermoplastic composites with flat energy directors and its application to the definition of optimum processing parameters. *Compos. Part A Appl. Sci. Manuf.* **2014** *65*, 27–37. [CrossRef]
22. Palardy, G.; Fernandez Villegas, I. Ultrasonic welding of thermoplastic composites with at energy directors: Influence of the thickness of the energy director on the welding process. In Proceedings of the 20th International Conference on Composite Materials, Copenhagen, Denmark, 19–24 July 2015; pp. 1–11.
23. Fernandez Villegas, I. In situ monitoring of ultrasonic welding of thermoplastic composites through power and displacement data. *J. Thermoplast. Compos. Mater.* **2015**, *28*, 66–85. [CrossRef]
24. Fernandez Villegas, I.; Valle Grande, B.; Bersee, H.E.N.; Benedictus, R. A comparative evaluation between flat and traditional energy directors for ultrasonic welding of CF/PPS thermoplastic composites. *Compos. Interfaces* **2015**, *22*, 717–729. [CrossRef]
25. Senders, F.J.M. Continuous Ultrasonic Welding of Thermoplastic Composites. Master's Thesis, Delft University of Technology, Delft, The Netherlands, 2016.
26. Benatar, A.; Gutowski, T.G. Ultrasonic welding of PEEK graphite APC-2 composites. *Polym. Eng. Sci.* **1989**, *29*, 1705–1721 [CrossRef]
27. Bonmatin, M.; Chabert, F.; Bernhart, G.; Cutard, T.; Djilali, T. Ultrasonic welding of CF/PEEK composites: Influence of welding parameters on interfacial temperature profiles and mechanical properties. *Compos. Part A Appl. Sci. Manuf.* **2022**, *162*, 107074. [CrossRef]
28. Mohan, S. Energy Directors for Continuous Ultrasonic Welding of Thermoplastic Composites. Master's Thesis, Delft University of Technology, Delft, The Netherlands, 2019.
29. Jongbloed, B.C.P. Sequential Ultrasonic Welding of Thermoplastic Composites. Master's Thesis, Delft University of Technology, Delft, The Netherlands, 2016.
30. Williams, G.; Green, S.; McAfee, J.; Heward, C.M. Induction welding of thermoplastic composites. In Proceedings of the Fibre Reinforced Composites, Liverpool, UK, 27–29 March 1990; pp. 133–136.
31. Cogswell, F.N. Thermoplastic interlayer bonding for aromatic polymer composites. In Proceedings of the 34th International SAMPE Symposium, Reno, NV, USA, 8–11 May 1989; pp. 2315–2325.
32. Border, J.; Salas, R. Induction heated joining of thermoplastic composites without metal susceptors. In Proceedings of the 34th International SAMPE Symposium, Reno, NV, USA, 8–11 May 1989; pp. 2569–2578.
33. Amanat, N.; Chaminade, C.; Grace, J.; McKenzie, D.R.; James, N.L. Transmission laser welding of amorphous and semi-crystalline poly-ether–ether–ketone for applications in the medical device industry. *Mater. Des.* **2010**, *31*, 4823–4830. [CrossRef]
34. Torrisi, L.; Caridi, F.; Visco, A.M.; Campo, N. Polyethylene welding by pulsed visible laser irradiation. *Appl. Surf. Sci.* **2011**, *257*, 2567–2575. [CrossRef]
35. Martineau, L.; Chabert, F.; Boniface, B.; Bernhart, G. Effect of interfacial crystalline growth on autohesion of PEEK. *Int. J. Adhes. Adhes.* **2019**, *89*, 82–87. [CrossRef]
36. Hsiao, B.S.; Sauer, B.B. Glass transition, crystallization, and morphology relationships in miscible poly(aryl ether ketones) and poly(ether imide) blends. *J. Polym. Sci. B Polym. Phys.* **1993**, *31*, 901–915. [CrossRef]
37. Fernandez Villegas, I.; Vizcaino Rubio, P. On avoiding thermal degradation during welding of high-performance thermoplastic composites to thermoset composites. *Compos. Part A Appl. Sci. Manuf.* **2015**, *77*, 172–180. [CrossRef]
38. Potente, H.; Becker, F.; Fiegler, G.; Korte, J. Investigations towards application of a new technique on laser transmission welding. *Weld. World* **2001**, *45*, 15–20.
39. Audoit, J.; Rivière, L.; Dandurand, J.; Lonjon, A.; Dantras, E.; Lacabanne, C. Thermal, mechanical and dielectric behaviour of poly(aryl ether ketone) with low melting temperature. *J. Therm. Anal. Calorim.* **2019**, *135*, 2147–2157. [CrossRef]
40. Merdas, I.; Thominette, F.; Verdu, J. Humid aging of polyetherimide. I. Water sorption characteristics. *J. Appl. Polym. Sci.* **2000**, *77*, 1439–1444. [CrossRef]
41. Grayson, M.A.; Wolf, C.J. The solubility and diffusion of water in poly(aryl-ether-ether-ketone) (PEEK)*. *J. Polym. Sci. B Polym. Phys.* **1997**, *25*, 31–41. [CrossRef]
42. Correa, A.; De Nicola, A.; Scherillo, G.; Loianno, V.; Mallamace, D.; Mallamace, F.; Ito, H.; Musto, P.; Mensitieri, G. A molecular interpretation of the dynamics of diffusive mass transport of water within a glassy polyetherimide. *Int. J. Mol. Sci.* **2021**, *22*, 2908. [CrossRef]
43. Augh, L.; Gillespie, J.W.; Fink, B.K. Degradation of continuous carbon fiber reinforced polyetherimide composites during induction heating. *J. Thermoplast. Compos. Mater.* **2001**, *14*, 96–115. [CrossRef]
44. Amancio-Filho, S.T.; Roeder, J.; Nunes, S.P.; dos Santos, J.F.; Beckmann, F. Thermal degradation of polyetherimide joined by friction riveting (FricRiveting). Part I: Influence of rotation speed. *Polym. Degrad. Stab.* **2008**, *93*, 1529–1538. [CrossRef]
45. Doumeng, M.; Makhlouf, L.; Berthet, F.; Marsan, O.; Delbé, K.; Denape, J.; Chabert, F. A comparative study of the crystallinity of polyetheretherketone by using density, DSC, XRD, and Raman spectroscopy techniques. *Polym Test.* **2021**, *93*, 106878. [CrossRef]
46. Ren, T.; Zhu, G.; Zhang, C.; Hou, X. Improving the interface of carbon-fibre-reinforced poly(ether-ether-ketone) composites withfibre surface coating: A review. *Polym Int.* **2022**, *71*, 741–750. [CrossRef]
47. de Almeida, O.; Bessard, E.; Bernhart, G. DSC investigation of the influence of carbon content on peek crystallisation and stability. In Proceedings of the 19th International Conference on Composite Materials, Montreal, QC, Canada, 28 July–2 August 2013; pp. 1–5.

48. Gaitanelis, D.; Worrall, C.; Kazilas, M. Detecting, characterising and assessing PEEK's and CF-PEEK's thermal degradation in rapid high-temperature processing. *Polym. Degrad. Stab.* **2022**, *204*, 110096. [CrossRef]
49. Balquet, A. *Méthodes D'assemblage de Thermoplastiques Renforcés Fibres Courtes ou Longues*; Rapport 213834R; ISPA Entreprise, Normandie Aerospace: Damigny, France, 2015; pp. 1–101.
50. Coulson, M.; Dantras, E.; Olivier, P.; Gleizes, N.; Lacabanne, C. Thermal conductivity and diffusivity of carbon-reinforced polyetherketoneketone composites. *J. Appl. Polym. Sci.* **2019**, *136*, 47975. [CrossRef]
51. Rudolf, R.; Mitschang, P.; Neitzel, M. Induction heating of continuous carbon-fibre-reinforced thermoplastics. *Compos. Part A Appl. Sci. Manuf.* **2000**, *31*, 1191–1202. [CrossRef]
52. Awaja, F.; Zhang, S. Self-bonding of PEEK for active medical implants applications. *J. Adhes. Sci. Technol.* **2015**, *29*, 1593–1606. [CrossRef]
53. Awaja, F.; Zhang, S.; McKenzie, D.R. Autohesion of semi-crystalline PEEK near and under the glass transition temperature. *Appl. Surf. Sci.* **2013**, *282*, 571–577. [CrossRef]
54. Mayboudi, L.S. Heat Transfer Modelling and Thermal Imaging Experiments in Laser Transmission Welding of Thermoplastics. Ph.D. Thesis, Queen's University, Kingston, ON, Canada, 2008.
55. Köhler, F.; Fernandez Villegas, I.; Dransfeld, C.; Herrmann, A. Static ultrasonic welding of carbon fibre unidirectional thermoplastic materials and the influence of heat generation and heat transfer. *J. Compos. Mater.* **2021**, *55*, 2087–2102. [CrossRef]
56. Ilie, M.; Kneip, J.C.; Matteï, S.; Nichici, A.; Roze, C.; Girasole, T. Through-transmission laser welding of polymers–temperature field modeling and infrared investigation. *Infrared Phys. Technol.* **2007**, *51*, 73–79. [CrossRef]
57. Speka, M.; Matteï, S.; Pilloz, M.; Ilie, M. The infrared thermography control of the laser welding of amorphous polymers. *Nondestruct. Test. Evaluation.* **2008**, *41*, 178–183. [CrossRef]
58. Akué Asséko, A.C.; Cosson, B.; Schmidt, F.; Maoult, Y.L.; Lafranche, E. Laser transmission welding of composites. Part A: Thermo-physical and optical characterization of materials. *Infrared Phys. Technol.* **2015**, *72*, 293–299. [CrossRef]
59. Akué Asséko, A.C.; Cosson, B.; Schmidt, F.; Le Maoult, Y.; Gilblas, R.; Lafranche, E. Laser transmission welding of composites. Part B: Experimental validation of numerical model. *Infrared Phys. Technol.* **2015**, *73*, 304–311. [CrossRef]
60. Gunawardena, D.S.; Cheng, X.; Cui, J.; Edbert, G.; Lu, L.; Ho, Y.T.; Tam, H.Y. Regenerated polymer optical fiber Bragg gratings with thermal treatment for high temperature measurements. *Photonics Res.* **2022**, *10*, 1011. [CrossRef]
61. Kisaka, P.A.; Beres-Pawlik, E.M.; Wojcik, J.; Wojcik, W. Fiber Bragg grating sensors for temperature measurement. In *Lightguides and Their Applications II*; SPIE: Bellingham, WA, USA, 2004; Volume 5576, pp. 270–276.
62. Ambrosio, D.; Aldanondo, E.; Wagner, V.; Dessein, G.; Garnier, C.; Vivas, J.; Cahuc, O. A semi-empirical model for peak temperature estimation in friction stir welding of aluminium alloys. *Sci. Technol. Weld. Join.* **2022**, *27*, 491–500. [CrossRef]
63. Baldan, A. Adhesively-bonded joints in metallic alloys, polymers and composite materials: Mechanical and environmental durability performance. *J. Mater. Sci.* **2004**, *39*, 4729–4797. [CrossRef]
64. Adams, R.D.; Comyn, J.; Wake, W.C. *Structural Adhesive Joints in Engineering*, 2nd ed.; Springer Science & Business Media: Berlin/Heidelberg, Germany; Chapman & Hall: London, UK, 1997.

materials

MDPI

Article

Investigating the Effect of Interface Temperature on Molecular Interdiffusion during Laser Transmission Welding of 3D-Printed Composite Parts

Anh-Duc Le, André Chateau Akué Asséko *, Benoît Cosson and Patricia Krawczak *

IMT Nord Europe, Institut Mines Télécom, University of Lille, Centre for Materials and Processes, F-59653 Villeneuve d'Ascq, France; anh-duc.le@imt-nord-europe.fr (A.-D.L.); benoit.cosson@imt-nord-europe.fr (B.C.)
* Correspondence: andre.akue.asseko@imt-nord-europe.fr (A.C.A.A.); patricia.krawczak@imt-nord-europe.fr (P.K.)

Abstract: The present study investigated the influence of temperature on molecular interdiffusion at the interface during the laser transmission welding of 3D-printed continuous carbon-fiber-reinforced thermoplastic composites. In order to accurately measure the temperature at the weld interface, a series of thermocouples were embedded in the laser-absorbent composite part. Two different molecular interdiffusion models were implemented to calculate the degree of healing and to predict the effects of temperature on the welding process. The degree of healing and the weld line width were computed and compared with microscopy observations. The discrepancy between the two proposed numerical models was less than 6%. Both models showed good agreement with the experimental data, with an average error of 13.28% and 7.26%, respectively. The results revealed a significant correlation between the thermal history and molecular interdiffusion at the interface. Furthermore, the relationship between the welding parameters (laser beam scanning speed) and weld line width was established. The findings of this study provide a comprehensive understanding of the underlying mechanisms involved in the laser welding of 3D-printed composites and offer insights to optimize the welding process for enhanced weld quality and superior mechanical properties in the final product.

Keywords: molecular interdiffusion; laser transmission welding; 3D printing; composite materials; thermoplastic polymers; interface temperature; modeling; simulation

Citation: Le, A.-D.; Akué Asséko, A.C.; Cosson, B.; Krawczak, P. Investigating the Effect of Interface Temperature on Molecular Interdiffusion during Laser Transmission Welding of 3D-Printed Composite Parts. *Materials* **2023**, *16*, 6121. https://doi.org/10.3390/ma16186121

Academic Editor: Ricardo J. C. Carbas

Received: 15 August 2023
Revised: 5 September 2023
Accepted: 6 September 2023
Published: 7 September 2023

Copyright: © 2023 by the authors. Licensee MDPI, Basel, Switzerland. This article is an open access article distributed under the terms and conditions of the Creative Commons Attribution (CC BY) license (https://creativecommons.org/licenses/by/4.0/).

1. Introduction

Laser transmission welding (LTW) presents numerous advantages in welding thermoplastic polymers and composites compared to alternative conventional techniques such as ultrasonic or friction welding. LTW offers precision, flexibility, a limited heat-affected zone, effortless automation and control, and the absence of contamination [1,2]. Regarding 3D-printed continuous fiber-reinforced composites, wherein the mechanical performance remains restricted by the fiber orientation within the printing layers, and the relatively low strength of inter-layer bonding [3–5], LTW emerges as a prospective joining technique to form large functional assemblies. It makes it possible to produce objects in which the continuous reinforcing fibers are arranged to support multi-directional mechanical loads.

In LTW, a pair of components are involved, wherein one exhibits transparency to the laser wavelength (so-called laser-transparent part) while the other possesses absorption characteristics at the same wavelength (so-called laser-absorbent part). Prior to initiating the welding process, these components are accurately aligned. Subsequently, the laser beam energy penetrates through the transparent material and is absorbed by the absorbent material at the interface of both materials. As a consequence, the absorbent interface experiences heating, causing the materials to melt and fuse together and creating a bond between the two parts. The quality of the bond formed by a weld line is characterized by two main phenomena that define the degree of bonding: intimate contact and healing (or the molecular interdiffusion of the polymer across the bonding interface) [6]. The phenomenon of healing is intricately linked with the process of intimate contact, as it is only possible for healing to take place within regions of the interface where intimate contact is established [7]. The present study focuses on the description of the healing process assuming that complete intimate contact is already achieved at the interface. Healing can occur only when the temperature rises above the glass transition temperature (T_g) for amorphous thermoplastics or the melt temperature (T_m) for semi-crystalline thermoplastics, respectively. For the welding of fiber-reinforced thermoplastic composites, it was demonstrated in a previous work that the temperature at the interface should be well above the (T_g) of the thermoplastics to achieve optimal adhesion and complete chain entanglement [8].

The establishment of a robust bond between welded components depends on both an optimal temperature distribution within the heat-affected zone and a controlled welding time [9]. The temperature distribution and weld profile at the interface were widely determined by numerical simulations [10–13]. Validation was achieved indirectly through infrared thermography measurement at the back surface of the absorbent part [8,14]. Nevertheless, numerical simulations require knowledge of the intensity flux at the weld interface, which depends on the intensity flux distribution of the incident laser beam as well as the thickness, structure, and optical properties of the transparent part. Unfortunately, in the case of fiber-reinforced and filled composites, or semi-crystalline polymers, measuring the energy loss due to reflection and absorption within the transparent part is not trivial [15].

Recent literature reviews have highlighted the growing significance of thermoplastic composite bonding and welding processes in various industries, particularly due to the demand for lightweight, high-strength components [16–18]. However, challenges persist in achieving efficient and reliable welds, especially when dealing with complex geometries and various polymer materials [16,17]. The main issues related to the welding process of thermoplastic composites include (1) the need to control the heat input to avoid burning the materials; (2) the need to ensure that the materials are properly bonded together at the interface; and (3) the need to prevent the formation of voids or defects in the weld. This research endeavors to address these challenges by exploring innovative solutions to improve the thermoplastic composite welding process [19–21]. By examining the effect of temperature on molecular interdiffusion at the interface, this study aims to offer insights into optimizing the welding parameters, such as the laser beam scanning speed, for enhanced weld strength and quality of the final product.

Previous studies in the literature have shown that temperature plays a crucial role in controlling healing at polymer–polymer interfaces [7,22,23]. As a consequence, in this study, we aim to comprehensively investigate the effect of temperature on molecular interdiffusion at the interface in the laser transmission welding of 3D-printed continuous carbon-fiber-reinforced thermoplastic composites. The goal is to gain a better understanding of the fundamental processes that occur during the laser welding of 3D-printed thermoplastic composites. This knowledge can be used to optimize the welding process and improve the quality and mechanical properties of the final product. A unique approach in this research work is the use of an embedded set of thermocouples in the laser-absorbent composite parts, which enables an accurate temperature measurement at the weld interface. Two different non-isothermal molecular interdiffusion models are employed to predict the effects of temperature on the welding process so as to investigate the relationship between temperature and molecular interdiffusion at the interface. As mentioned above, intimate contact issues are beyond the scope of the present study; therefore, weld quality is characterized by the weld line width instead of the weld strength [24,25]. Accordingly, the relationship between the welding parameters and weld line width is analyzed.

2. Molecular Interdiffusion Models

In the case of isothermal conditions, the reptation theory introduced by De Gennes [26] and Doi and Edwards [27], which models the motion of individual linear polymer chains within the amorphous bulk, is frequently employed to describe the molecular interdiffusion (or healing) process. In the model, a polymer chain of length L is considered to be confined in an imaginary tube, which is defined by the constraints of neighboring polymer chains via entanglements. This imaginary tube restricts the movement of the chain exclusively along its curvilinear length. At the beginning of the process, $t = 0$, the chain (i.e., the thin solid line in Figure 1) is totally encompassed by the original tube. After a period of time, $t = t_1$, the end of the chain, which has more mobility compared to the center of mass of the chain, escapes from the original tube, forming "minor chains" of length l. As time evolves, an increasing proportion of the chain leaves the original tube. During this process, l increases with time until it reaches ($l = L$) at the reptation time t_R (see Figure 1).

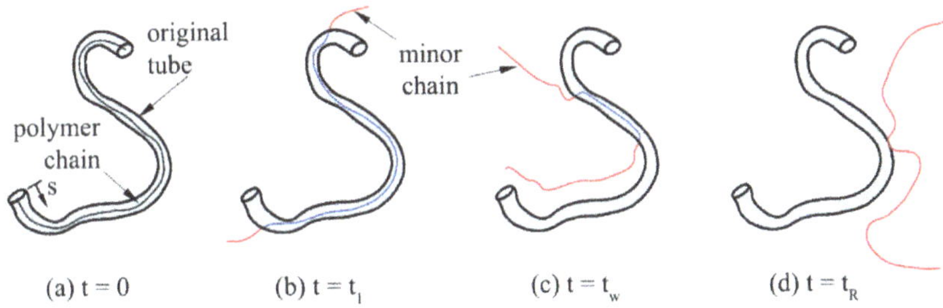

Figure 1. Reptation movement of a linear polymer chain [28].

The polymer molecular interdiffusion at the interface between two miscible thermoplastics during the LTW process is depicted schematically in Figure 2. At the origin ($t = 0$), all the minor chains have zero length as denoted by the dots. At $t = t_1$, the reptation of the minor chains initiates, and some of the minor chains cross the interface by an average distance denoted as χ. Finally, when the time is close to the reptation time or relaxation time $t = t_R$, the interdiffusion is fully developed and chain entanglements are achieved (i.e., $\chi = \chi_\infty$, and $l = L$) [22,29]. For a time longer than the reptation time, the interface disappears, and the properties of the virgin bulk material are reached [30]. The bond strength, σ, is characterized by the average interpenetration distance of the chains across the interface χ, which varies as the square root of the minor chain length as $\chi \approx \sqrt{l}$. By using Einstein's diffusion equation (Equation (1)), it is possible to show that healing is related to time to a power of 1/4 [31]:

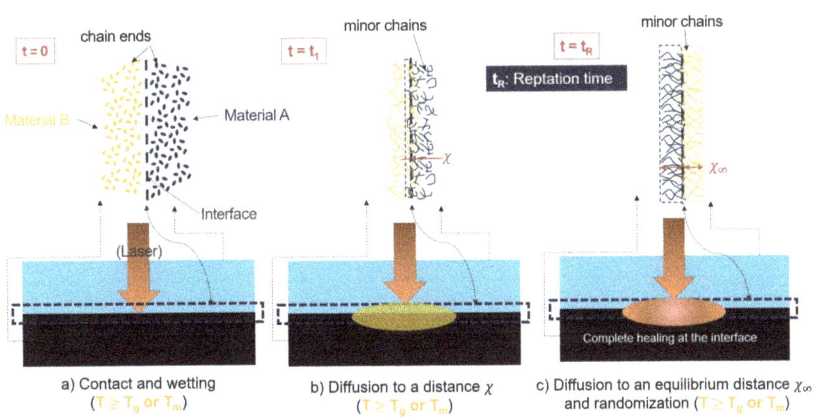

Figure 2. Interdiffusion of minor chains across a polymer–polymer interface during the laser transmission welding process.

$$\left\langle l^2 \right\rangle = 2Dt \Rightarrow \chi \approx \sqrt{l} = \left\langle l^2 \right\rangle^{1/4} \propto t^{1/4} \tag{1}$$

where $\left\langle l^2 \right\rangle$ is the quadratic distance of diffusion, and D is the self-diffusion coefficient Finally, the degree of healing, D_h, is defined as the ratio of the instantaneous interfacial bond strength to the ultimate bond strength as [7,26]

$$D_h = \frac{\sigma}{\sigma_\infty} = \frac{\chi}{\chi_\infty} = \left(\frac{l}{L}\right)^{1/2} = \left(\frac{t}{t_R}\right)^{1/4} \tag{2}$$

Regarding the LTW process, the temperature field exhibits temporal variation. Furthermore, the temperature distribution within a weld line is uneven. Bastien and Gillespie [23] were the pioneers in introducing the non-isothermal healing model for the fusion bonding of amorphous polymers. In this model, the thermal history can be divided into n time intervals, in which the average temperature T_i^*, between times t_i and t_{i+1}, is assumed constant. In this manner, the application of the healing theory is feasible during every individual isothermal increment. Bastien and Gillespie [23] proposed two mathematical models for the prediction of the bond strength, which were based on the minor chain length criteria and the average interpenetration distance criteria. Szuchács et al. [28] have recently demonstrated that the first model (Equation (3)) is more appropriate compared to their experimental data. Therefore, this model (hereafter "Model 1") is employed in the framework of this research work:

$$D_h = \frac{\sigma}{\sigma_\infty} = \left(\frac{l}{L}\right)^{1/2} = \left[\sum_{i=0}^{n}\left(\frac{t_{i+1}^{1/2} - t_i^{1/2}}{t_R(T_i^*)^{1/2}}\right)\right]^{1/2} \tag{3}$$

The bond strength can also be assessed by calculating the quadratic distance of diffusion according to the formulation depicted in Equation (1). On the basis of the work of Doi and Edwards [32], Graessley [33] pointed out that the self-diffusion coefficient may be calculated from the measurable viscoelastic quantities of the polymer as in Equation (4):

$$D = \frac{G_N^0}{135}\left(\frac{\rho RT}{G_N^0}\right)^2\left(\frac{\langle r^2\rangle}{M_w}\right)\frac{M_c(T)}{M_w^2\eta_{0,M_c(T)}} \tag{4}$$

where G_N^0 is the plateau modulus, ρ is the polymer density, R is the universal gas constant, T is the absolute temperature, $\langle r^2\rangle$ is the unperturbed mean square end-to-end distance of the chain, M_w is the molecular weight, $M_c(T)$ is the critical molecular weight at the temperature T, and $\eta_{0,M_c(T)}$ is the zero-shear viscosity at the critical molecular weight and the temperature T.

The G_N^0 modulus can be deduced from the master curve of the considered polymer, which can be recorded through dynamic rheological measurements. The critical molecular weight $M_c(T)$ is linked to the molecular weight of entanglements $M_e(T)$ by Equation (5) [34]:

$$M_c(T) = 2M_e(T) \quad \text{with} \quad M_e(T) = \frac{\rho RT}{G_N^0} \tag{5}$$

The unperturbed mean square end-to-end distance of the chain is difficult to determine experimentally. It was stated that the ratio between this distance and the molar mass M_w is a constant [35]. In the case of the thermoplastic polymer used in the present study, namely poly-ethylene terephthalate glycol-modified (PETG), this ratio is fixed as [35]

$$\frac{\langle r^2\rangle}{M_w} = 0.845\overset{o}{A}{}^2\,\mathrm{mol{\cdot}g}^{-1} \tag{6}$$

Note that the value associated with PET was used for this calculation due to the overall similarities and a lack of prior literature covering PETG diffusion dynamics.

The relationship between the zero-shear viscosity η_0 and the molecular weight M_w can be expressed by the following equation [33]:

$$\eta_0 = k_H M_w^{3.4} \quad (7)$$

where k_H is a material constant. Utilizing the values for the known zero-shear viscosity and molecular weight in Ref. [36], it is possible to solve for k_H. Subsequently, a molecular weight of $M_w = 28{,}126$ g·mol^{-1} was estimated for the PETG used in the present study.

Since the numerical value for $\eta_{0,M_c(T)}$ was not available, we used the following relationship (Equation (8)) for the calculations, as also carried out by Kim and Han [37]:

$$\frac{\eta_{0,M_c(T)}}{\eta_{0,M_w(T)}} = \left(\frac{M_c(T)}{M_w(T)}\right)^{3.4} \quad (8)$$

As suggested in [8,38,39], complete interdiffusion can only be achieved when the quadratic distance of diffusion is greater than the mean square end-to-end distance ($\langle r^2 \rangle = 2.38 \times 10^{-16}$ m^2). Therefore, the degree of healing (D_h) can be calculated by the following equation (hereafter "Model 2"):

$$D_h = \frac{\sigma}{\sigma_\infty} = \left(\frac{\langle l^2 \rangle}{\langle r^2 \rangle}\right)^{1/4} = \left[\sum_{i=0}^{n} \frac{2D(T_i^*)(t_{i+1} - t_i)}{\langle r^2 \rangle}\right]^{1/4} \quad (9)$$

3. Experimentals

3.1. Materials

The 3D-printed specimens used for the welding experiments were made of polyethylene terephthalate glycol-modified (PETG) and continuous carbon-fiber-reinforced PETG composites (CCFPC). A natural transparent and a black-pigmented PETGs were chosen for the semi-transparent and the absorbent parts, respectively. The materials were supplied by Polymaker™ in the form of 1.75 mm round filaments. According to the supplier, the PETG polymer had a density of 1.25 g·cm^{-3} and a glass transition temperature of 81 °C. The nozzle temperature was recommended in the range of (230–260 °C). The composite carbon fiber (CCF) filament of 0.35 mm diameter was supplied by Anisoprint™, which comprised 60% in volume of carbon fibers pre-impregnated with an epoxy thermosetting resin to provide good adhesion with the thermoplastic matrix.

3.2. 3D Printing of Continuous Carbon-Fiber-Reinforced Thermoplastic Composites

The CCFPCs were printed using a Composer A4 Desktop 3D Printer from Anisoprint™. The printing process was based on the composite fiber co-extrusion technology. In this printing technology, both the polymer and the CCF filaments are fed from two different spools to a common printing head (see Figure 3). In this way, the molten polymer wets the CCF in the nozzle prior to printing. The Aura™ software was used to slice the 3D CAD models and assign the processing parameters to the G-codes. The 3D CAD models used to manufacture the welding specimens had a rectangular bar shape of 80 × 20 × 2 mm^3. The nozzle temperature was T_n = 260 °C, and the bed temperature was T_b = 75 °C. The printing speed was 10 mm·s^{-1} and the layer thickness was 0.2 mm for the plastic nozzle and 0.4 mm for the composite nozzle. The fill density was 100%, and the fiber fill type was set to line with an angle of 0°.

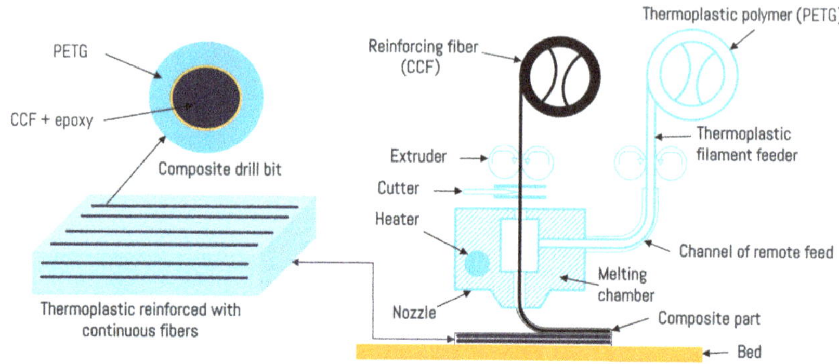

Figure 3. Schematic view of the Anisoprint™ printer used to manufacture the continuous carbon-fiber-reinforced PETG composites.

3.3. Laser Transmission Welding of Composites

The LTW experiments were conducted using two types of 3D-printed components: natural PETG-T, which is semi-transparent at the laser wavelength, and CCFPC-A consisting of black-pigmented PETG and CCF, which is absorbent at the same wavelength. The process setup is illustrated in Figure 4. The laser welding machine (LEISTER NOVOLAS™) was equipped with a diode laser transported by an optic fiber. The maximum output power was 46 W and the laser wavelength was 0.940 μm. The semi-transparent PETG-T part was placed above the CCFPC-A absorbent part on a workbench and fixed with a transparent tape. The two components to be welded were located between a transparent cover and the workbench, which was displaced by a pneumatic actuator. The quartz glass used as the transparent cover allowed 95% of laser energy transmitted through it at the laser wavelength. For this application, the used laser power was set at 10 W in the welding system, and the clamping pressure was 6 bar in the pneumatic system. Two sequences of scan lines were run at different scanning speeds (i.e., v = 1.36 mm·s^{-1}, v = 2.72 mm·s^{-1}, and v = 4.08 mm·s^{-1}).

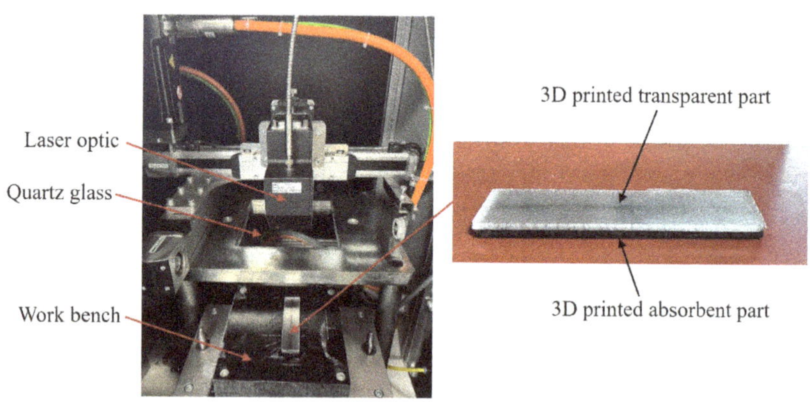

Figure 4. Laser transmission welding of 3D-printed components.

3.4. Temperature Measurement at the Weld Interface

The monitoring of the thermal evolution occurring at the interface during the LTW procedure was carried out using 5 k-type thermocouples with a wire diameter of 0.12 mm. The thermocouples were strategically embedded into designated locations (as specified in Table 1) within the holes of the absorbent component, with a penetration depth slightly exceeding the thermocouple junction size from the top surface. Once the thermocouples were appropriately positioned, the exact locations of the thermocouple junctions were reassessed. The measurements were conducted with accuracy of 0.01 mm (in Table 1). The samples were prepared with wide welding in order to create perfect contact. Following the initial trial, the thermocouples were completely wet in the thermoplastic material. Subsequently, repeated experiments were conducted on the same sample to measure the temperature field. The data were recorded using the Datataker DT85M data logger device (as illustrated in Figure 5). The frame rate for capturing the temperature at the interface was 120 Hz, and the measurement uncertainty was ±2.5 °C.

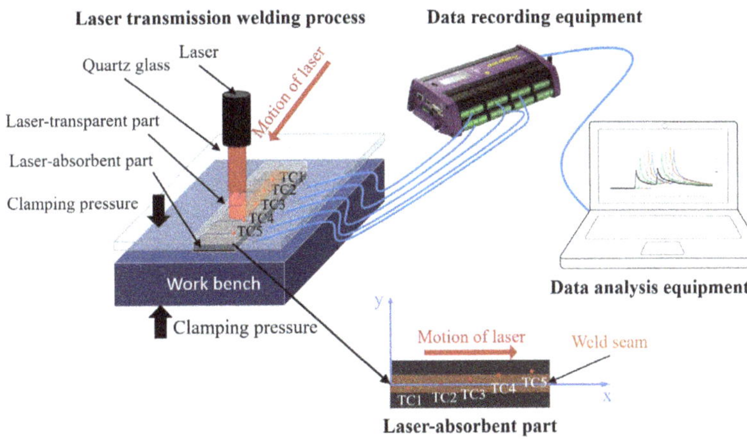

Figure 5. Illustration of laser transmission welding process and temperature measurement at the weld interface. TC stands for thermocouple.

Table 1. Coordinate positions of the thermocouples.

Thermocouple Number	TC1	TC2	TC3	TC4	TC5
x (mm)	10.00	24.00	37.95	52.80	67.00
y (mm)	−0.90	0.10	1.10	2.10	3.10

3.5. Rheological Characterization of the Thermoplastic

The rheological properties of PETG were investigated using an Anton Paar Physica MCR 301 rotational rheometer. The measurements were conducted using a parallel-plate geometry with a diameter of 35 mm at a 1 mm gap. Nitrogen was employed as a protective gas to prevent polymer degradation. The rheological measurement specimens were 3D printed, employing identical parameters to those employed for the production of the PETG-T parts. Frequency sweep tests were performed to determine the storage modulus G', the loss modulus G'' and the complex viscosity η^* at different temperatures: 130 °C, 150 °C, 170 °C, 190 °C, 210 °C and 230 °C. With respect to the linear viscoelastic range, the shear strain applied was 5%. The angular frequency was varied from 0.1 rad·s^{-1} to 628 rad·s^{-1}.

4. Results and Discussion

4.1. Temperature Measurement

Figure 6a illustrates the temporal variation in the interface temperature recorded by five thermocouples following two successive passages of the laser beam. The highest temperature was recorded by TC2 positioned at the center of the laser spot, while the temperatures gradually decreased with increasing distance from the central line of the part The cooling was instantaneous following the passage of the laser on each thermocouple via thermal conduction and natural convection. In comparison with our previous research conducted on thermoplastic PLA [40], the cooling rate was accelerated as a result of the higher thermal conductivity of the CCFPC compared to the pure thermoplastic PLA.

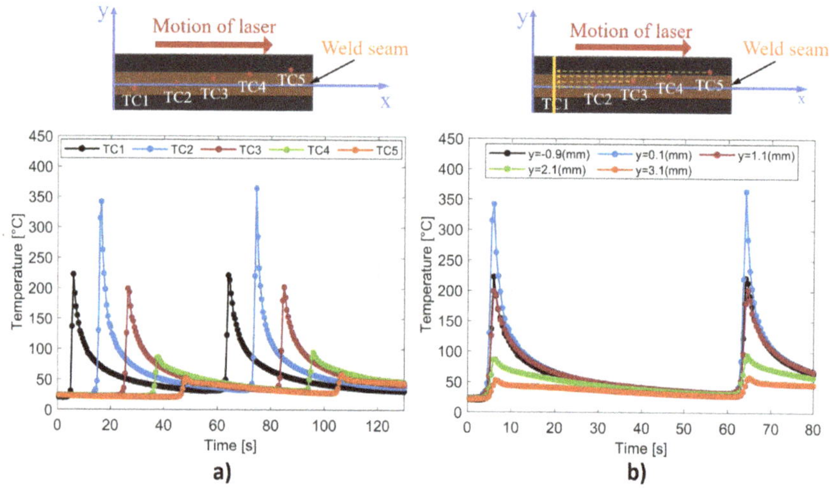

Figure 6. (**a**) Experimental interface temperatures measured at five different positions with five thermocouples. (**b**) Thermal history at five different positions on a perpendicular line passing through the thermocouple TC1 using shifted time.

For the LTW process with a constant scanning speed of the laser, except for the two end sections of the part subjected to boundary effects, the thermal history of all perpendicular lines to the weld seam was considered to be identical. Accordingly, the temperature measurement by five thermocouples at different positions along the weld seam could be used to describe the thermal history occurring on a single line perpendicular to the weld seam, as shown in Figure 6b. For this purpose, a shift factor was applied to the time, which corresponded to the duration taken by the laser to travel a distance equal to the spatial gap between the thermocouple position and the specified line.

The instantaneous temperature profile along a perpendicular line to the weld seam has been demonstrated to have a Gaussian shape [40,41]. Based on the outcomes presented above, it is evident that employing a Gaussian-fitted model makes it possible to generate a temperature profile for every temporal instance (see Figure 7a). Finally, the full thermal history of the specified line was computationally reconstructed, as shown in Figure 7b. These temperature data were then used for the calculation of the degree of healing.

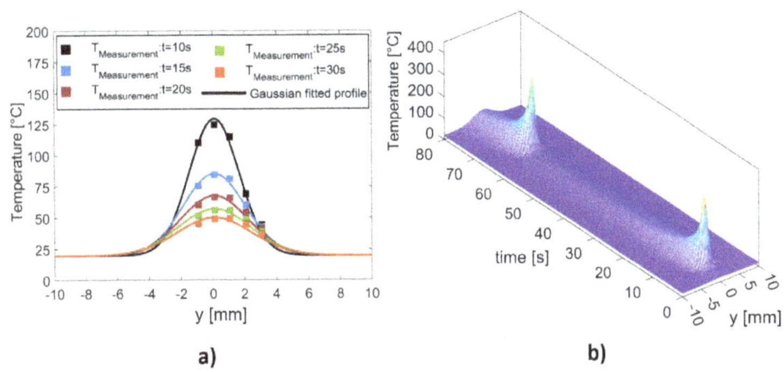

Figure 7. (**a**) Gaussian−fitted profile for temperature on a line perpendicular to the weld seam at different moments. (**b**) Full thermal history of a line perpendicular to the weld seam during the laser transmission welding process.

4.2. Rheological Characterization

Figure 8a plots the results of the complex viscosity versus the angular frequency, obtained from dynamic frequency sweep tests at temperatures ranging from 130 °C to 230 °C. The variation in the viscosity can be well fitted by the Carreau–Yasuda model (Equation (10)), where the zero-shear viscosity (η_0) and the relaxation time (λ) are produced (as shown in the Table 2)

$$\eta = \eta_0\left[1 + (\lambda\omega)^a\right]^{\frac{n-1}{a}} \tag{10}$$

where a stands for the width of the transition range between zero-shear viscosity and the power law regime, and n is the power law exponent. The Carreau–Yasuda regression parameters giving the best fits to the experimental data are listed in Table 2.

Table 2. Carreau–Yasuda regression parameters.

Temperature (°C)	η_0 (Pa.s)	λ (s)	a	n
130	1,043,395	1.2363	0.56	0.09
150	135,215	0.1682	0.60	0.10
170	26,704	0.0481	0.71	0.18
190	8346	0.0207	0.81	0.26
210	3548	0.0131	0.90	0.35
230	1627	0.0117	1.20	0.52

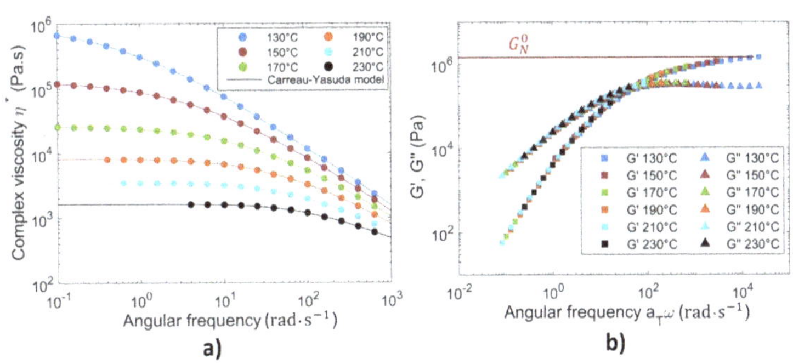

Figure 8. (**a**) Complex viscosity (circle symbol) versus angular frequency. (**b**) Time–temperature superposition master curves of storage modulus G' (square symbol) and loss modulus G'' (triangle symbol) at 170 °C for PETG.

The master curves were constructed by time–temperature superposition (TTS) from dynamic frequency sweep tests obtained at different temperatures. Figure 8b shows the master curves of the storage modulus G' and the loss modulus G'' as functions of the angular frequency obtained at the reference temperature of 170 °C. The results demonstrate the linear viscoelastic characteristics of the PETG material. As illustrated in Figure 8b, the plateau zone of the G' curve is clearly visible when the frequency is above 10^4 rad·s^{-1} within the superposed angular frequency range. The plateau modulus G_N^0 was determined as equal to the storage modulus G' at the angular frequency, where G'' reaches a minimum in the plateau zone [42].

The widely accepted Williams–Landel–Ferry (WLF) model was employed to characterize the temperature-dependent behavior of the zero–shear viscosity as in Equation (11):

$$\eta_0 = D_1 \exp\left[\frac{-A_1(T - T_g))}{A_2 + (T - T_g)}\right] \tag{11}$$

where D_1, A_1, A_2 are data-fitted parameters. The reptation time (t_R) of an amorphous material was considered to be equivalent to the relaxation time (λ) [6,28,29], which was derived from viscosity measurements (see Table 2). It obeys also a WLF law as in Equation (12):

$$t_R(T) = a_T t_R(T_{ref}) \quad \text{with} \quad \log a_T = \frac{-C_1(T - T_{ref}))}{C_2 + (T - T_{ref})} \tag{12}$$

where a_T is the shift factor, and C_1 and C_2 are constants. The data-fitted parameters for WLF models of the zero-shear viscosity (η_0) and the reptation time (t_R) at T_{ref} = 170 °C are listed in Table 3. The semilog plot of the experimentally obtained data and the corresponding WLF fitted curves are given in Figure 9.

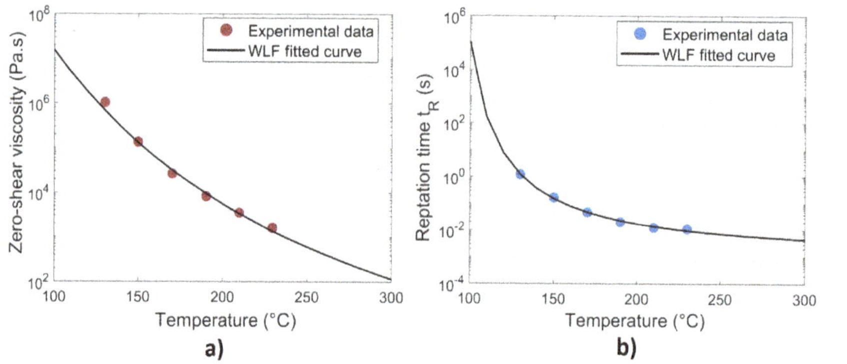

Figure 9. (**a**) Zero-shear viscosity versus temperature for PETG. (**b**) Reptation time versus temperature for PETG at T_{ref} = 170 °C.

Table 3. WLF fitted parameters for zero-shear viscosity (η_0) and the reptation time (t_R).

Parameter	D_1 (Pa.s)	A_1	A_2 (K)	C_1	C_2 (K)
Value	2.0115×10^8	25.65	171.94	1.7251	89

4.3. *Effects of Temperature on Molecular Interdiffusion at the Interface*

In order to investigate the effects of temperature on molecular interdiffusion at the interface, LTW trials with different scanning speeds were carried out: v = 1.36 mm·s^{-1}, v = 2.72 mm·s^{-1}, and v = 4.08 mm·s^{-1}. Temperature measurements were performed and the thermal histories of a line perpendicular to the weld seam were established, as depicted in Figure 10. In this figure, the glass transition temperature (T_g) is represented by the semi-transparent red plane. It appears that variations in the scanning speed yield disparities in the thermal history of the specified line. Decreasing the scanning speed leads to an enlarged weld zone where the temperature exceeds the glass transition temperature T_g, as well as an extended duration during which the temperature remains above T_g. These obtained data serve as essential variables for the calculation of the degree of healing along the specified line.

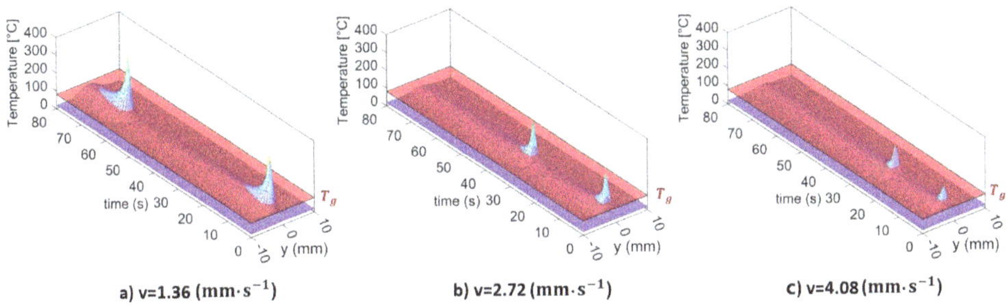

Figure 10. Thermal history of a line perpendicular to the weld seam for different scanning speeds of the laser beam.

Two different non-isothermal models, Equations (3) and (9), respectively, were implemented numerically to predict the degree of healing as a function of time. The calculations were performed for each node along the perpendicular line. It is important to note that the healing mechanism exclusively occurs when the interface temperature exceeds the glass transition temperature T_g. Therefore, for any time increment, if the temperature is below T_g, molecular interdiffusion will not take place at this node and the degree of healing will not increase. Once a degree of healing of unity is achieved, the state of full healing is preserved and remains unaffected by subsequent fluctuations in temperature (see Figure 11). For Model 1 (Equation (3)), only the reptation time (t_R) at given temperatures was needed for the calculation; it was calculated with the WLF equation (Equation (12) and Figure 9b). For Model 2 (Equation (9)), all the data for the self-diffusion coefficient calculation of the used PETG are reported in Table 4. The zero-shear viscosity was calculated using the WLF model (Equation (11) and Figure 9a). Then, the zero-shear viscosity at the critical molecular weight and the temperature T ($\eta_{0,M_c(T)}$) was computed according to Equation (8). The evolution of the self-diffusion coefficient (D) and the degree of healing (D_h) versus time, calculated at the location y = 0.8 mm for the case v = 2.72 mm·s^{-1}, is illustrated in Figure 11.

Table 4. Data for self-diffusion coefficient calculation of PETG.

Parameter	G_N^0 (Pa)	ρ (g·m^{-3})	R (J·(mol·K)$^{-1}$)	M_w (g·mol^{-1})	$\langle r^2 \rangle$ (m^2)
Value	1,298,300	1.25 × 10^6	8.314	28,126	2.38 × 10^{-16}

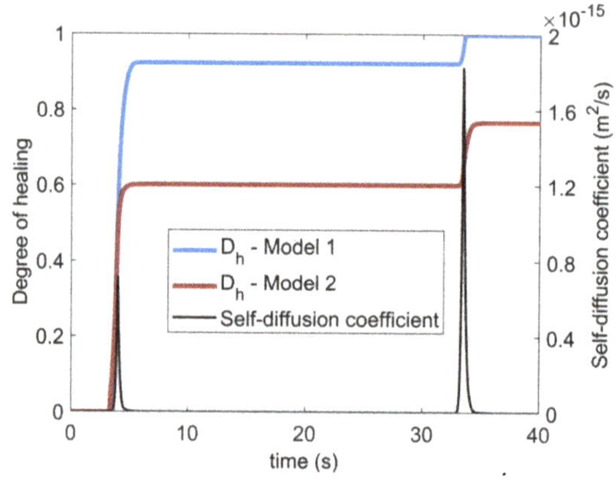

Figure 11. Evolution of self-diffusion coefficient (D) and degree of healing (D_h) versus time at $y = 0.8$ mm for scanning speed $v = 2.72$ mm·s^{-1}.

Figure 12 plots the calculated results obtained from the two non-isothermal models. It is obvious that there is a direct correlation between the decreased scanning speed and the increased width of the weld seam. This finding can be explained by the fact that the molecular interdiffusion is promoted by the elevated temperature at the interface and the longer welding duration. As a result, the lower the scanning speed is, the larger the weld line width is. It is evident that both models exhibit small disparities in their predictions of the degree of healing (D_h). For the first two cases with higher interface temperatures, complete healing is achieved at the center of the weld seam, wherein the degree of healing reaches unity ($D_h = 1$). Conversely, at the highest scanning speed, both models predict that complete healing has not yet been achieved within the weld seam.

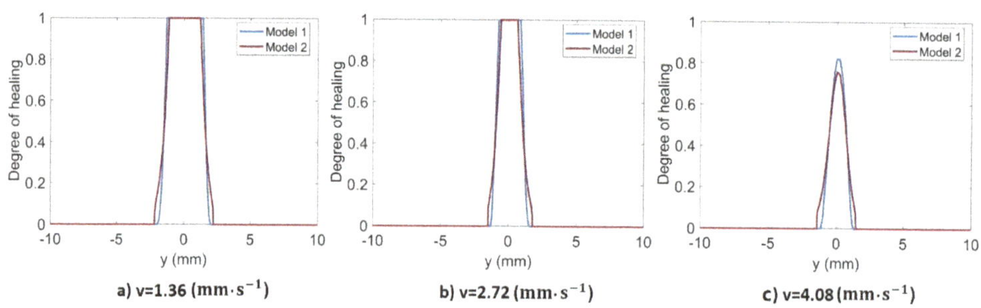

Figure 12. Computed degree of healing along a line perpendicular to the weld seam using two different models.

To assess the interdiffusion models, a comparative analysis was conducted between the predicted and experimental weld line widths. Optical microscopy examination of the cross-sectional assembled parts was performed to obtain information on the weld line width (W_{wl}), as depicted in Figure 13. For the visualization of the weld zone in the optical microscopy images, the degree of healing (D_h) at the edge of the weld seam should be relatively consistent. As reported by [24], D_h varies between 0 and 1 at the edge of the weld seam. For consistency, an average value of $D_h = 0.5$ was used in this study to predict the theoretical weld line widths for both numerical models. Figure 14 shows a comparison of the predicted and measured weld line widths as a function of the laser scanning speed. It appears that both models are reasonably efficient in predicting the weld line width and that the discrepancy between them is relatively small (i.e., less than 6%). The average errors of the two models compared to the experimental data are 13.28% and 7.26%, respectively.

Generally, Model 2 demonstrates a commendable predictive capability, which results in a better correlation with the experimental data across all investigated scenarios.

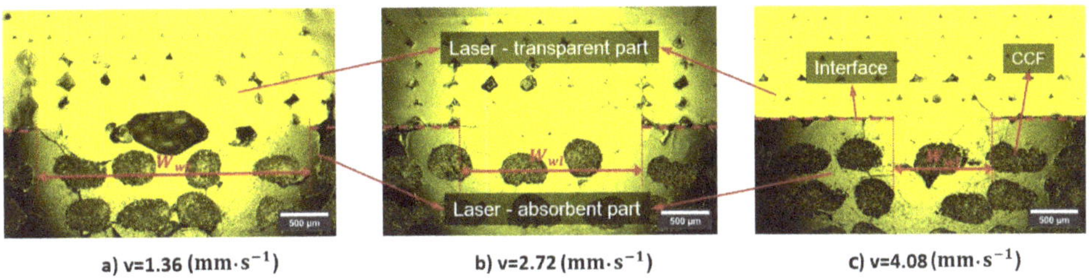

Figure 13. Computed weld line width using optical microscopy images.

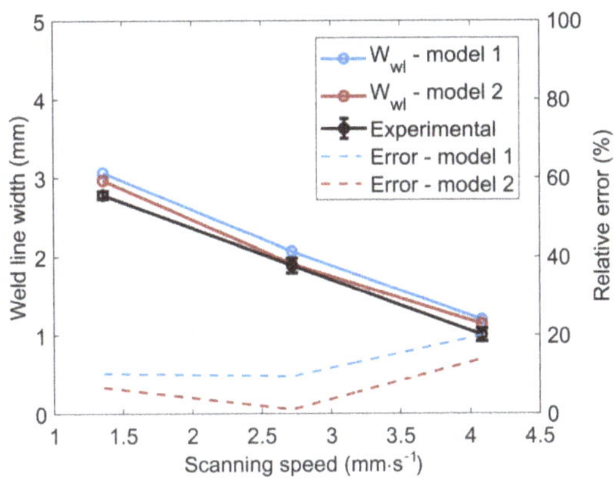

Figure 14. Prediction of weld line width as a function of laser scanning speed.

In Figure 13a, the sample was welded with excessively high line energy (i.e., the ratio of the laser power to the scanning speed); thereby, the interface was overheated, yielding thermal degradation and pores in the joint, as also reported in Ref. [9]. This experimental phenomenon could not be predicted with the proposed numerical methods.

5. Conclusions

A novel approach has been proposed for the comprehensive investigation of the effect of temperature on molecular interdiffusion during the laser transmission welding of 3D-printed continuous carbon-fiber-reinforced thermoplastic composites. The novelty of the approach lies in the implementation of an embedded set of thermocouples within the laser-absorbent component, which permits an accurate temperature measurement at the weld interface. Using the data obtained from five thermocouples embedded at different positions, the thermal history of a line perpendicular to the weld seam was reconstructed. Two non-isothermal molecular interdiffusion models were proposed and employed to predict the degree of healing across the specified line. The accuracy and reliability of the molecular interdiffusion models were assessed by comparing the weld line widths predicted by numerical models with the measurements derived from optical microscopy of the cross-sectional assembled parts. The results revealed a significant correlation between the thermal history and molecular interdiffusion at the weld interface. Moreover, the investigation highlighted the relationship between the welding parameters (i.e., the scanning speed) and weld line width. Both the proposed numerical models predicted the weld line width fairly accurately and produced a good correlation with the experimental data across all examined cases. The discrepancy between the two numerical modes was less than 6% and the average errors compared to the experimental data were 13.28% and 7.26%, respectively.

By understanding the mechanisms involved in the laser welding of 3D-printed composites, the findings can be utilized to optimize the welding process, resulting in improved weld quality and enhanced mechanical properties in the final product.

Author Contributions: Conceptualization, A.-D.L.; methodology, A.-D.L., A.C.A.A. and B.C.; software, A.-D.L.; validation, A.-D.L. and A.C.A.A.; formal analysis, A.-D.L.; investigation, A.-D.L. and A.C.A.A.; resources, A.-D.L.; data curation, A.-D.L. and A.C.A.A.; writing—original draft preparation, A.-D.L.; writing—review and editing, B.C., A.C.A.A. and P.K.; visualization, A.-D.L.; supervision, B.C., A.C.A.A. and P.K.; project administration, A.C.A.A. and P.K.; funding acquisition, A.C.A.A. All authors have read and agreed to the published version of the manuscript.

Funding: This research was funded by the National French Research Agency's ANR JCJC program through the AAPG 2021-CE10 "Industrie et Usine du Futur: Homme, Organisation, Technologies" initiative (grant agreement n° ANR-21-CE10-0007-01).

Institutional Review Board Statement: Not applicable.

Informed Consent Statement: Not applicable.

Data Availability Statement: The data presented in this study are available on request from the corresponding author.

Acknowledgments: The authors extend their sincere appreciation to the National French Research Agency's ANR JCJC program for funding the SHORYUKEN project (grant agreement n° ANR-21-CE10-0007-01) through the AAPG 2021-CE10 "Industrie et Usine du Futur: Homme, Organisation, Technologies" initiative. Without their support, this work would not have been possible.

Conflicts of Interest: The authors declare no conflict of interest.

References

1. Acherjee, B. Laser transmission welding of polymers—A review on process fundamentals, material attributes, weldability, and welding techniques. *J. Manuf. Process.* **2020**, *60*, 227–246. [CrossRef]
2. Wu, J.; Lu, S.; Wang, H.J.; Wang, Y.; Xia, F.B.; Wang, J. A review on laser transmission welding of thermoplastics. *Int. J. Adv. Manuf. Technol.* **2021**, *116*, 2093–2109. [CrossRef]
3. Chacón, J.M.; Caminero, M.A.; García-Plaza, E.; Núñez, P.J. Additive manufacturing of PLA structures using fused deposition modelling: Effect of process parameters on mechanical properties and their optimal selection. *Mater. Des.* **2017**, *124*, 143–157. [CrossRef]
4. Tian, X.; Liu, T.; Yang, C.; Wang, Q.; Li, D. Interface and performance of 3D printed continuous carbon fiber reinforced PLA composites. *Compos. Part Appl. Sci. Manuf.* **2016**, *88*, 198–205. [CrossRef]
5. Zhang, Z.; Yavas, D.; Liu, Q.; Wu, D. Effect of build orientation and raster pattern on the fracture behavior of carbon fiber reinforced polymer composites fabricated by additive manufacturing. *Addit. Manuf.* **2021**, *47*, 102204. [CrossRef]
6. Regnier, G.; Le Corre, S. Modeling of Thermoplastic Welding. In *Heat Transfer in Polymer Composite Materials*; John Wiley & Sons, Ltd.: Hoboken, NJ, USA, 2016; pp. 235–268. [CrossRef]
7. Yang, F.; Pitchumani, R. Healing of Thermoplastic Polymers at an Interface under Nonisothermal Conditions. *Macromolecules* **2002**, *35*, 3213–3224. [CrossRef]
8. Akué Asséko, A.C.; Cosson, B.; Lafranche, E.; Schmidt, F.; Le Maoult, Y. Effect of the developed temperature field on the molecular interdiffusion at the interface in infrared welding of polycarbonate composites. *Compos. Part B Eng.* **2016**, *97*, 53–61. [CrossRef]
9. Klein, R. *Laser Welding of Plastics: Materials, Processes and Industrial Applications*; John Wiley & Sons: Hoboken, NJ, USA, 2012.
10. Sooriyapiragasam, S.K.; Hopmann, C. Modeling of the heating process during the laser transmission welding of thermoplastics and calculation of the resulting stress distribution. *Weld. World* **2016**, *60*, 777–791. [CrossRef]
11. Aden, M. Influence of the Laser-Beam Distribution on the Seam Dimensions for Laser-Transmission Welding: A Simulative Approach. *Lasers Manuf. Mater. Process.* **2016**, *3*, 100–110. [CrossRef]
12. Chen, Z.; Huang, Y.; Han, F.; Tang, D. Numerical and experimental investigation on laser transmission welding of fiberglass-doped PP and ABS. *J. Manuf. Process.* **2018**, *31*, 1–8. [CrossRef]
13. Cosson, B.; Akué Asséko, A.C.; Lagardère, M.; Dauphin, M. 3D modeling of thermoplastic composites laser welding process—A ray tracing method coupled with finite element method. *Opt. Laser Technol.* **2019**, *119*, 105585. [CrossRef]
14. Ilie, M.; Kneip, J.C.; Matteï, S.; Nichici, A.; Roze, C.; Girasole, T. Through-transmission laser welding of polymers—Temperature field modeling and infrared investigation. *Infrared Phys. Technol.* **2007**, *51*, 73–79. [CrossRef]

15. Chen, M.; Zak, G.; Bates, P.J. Description of transmitted energy during laser transmission welding of polymers. *Weld. World* **2013**, *57*, 171–178. [CrossRef]
16. Enriquez, I.; Noronha, C.; Teo, K.; Sarmah, A.; Gupta, S.; Nandi, A.; Fishbeck, B.; Green, M.J.; Vashisth, A. Closed-Loop Composite Welding and Bonding System Using Radio-Frequency Heating and Pressure. *J. Compos. Sci.* **2023**, *7*, 116. [CrossRef]
17. Modi, V.; Bandaru, A.K.; Ramaswamy, K.; Kelly, C.; McCarthy, C.; Flanagan, T.; O'Higgins, R. Repair of Impacted Thermoplastic Composite Laminates Using Induction Welding. *Polymers* **2023**, *15*, 3238. [CrossRef]
18. Barazanchy, D.; van Tooren, M. Heating mechanisms in induction welding of thermoplastic composites. *J. Thermoplast. Compos. Mater.* **2023**, *36*, 473–492.
19. Jongbloed, B.C.P.; Teuwen, J.J.E.; Benedictus, R.; Villegas, I.F. A Study on Through-the-Thickness Heating in Continuous Ultrasonic Welding of Thermoplastic Composites. *Materials* **2021**, *14*, 6620. [CrossRef]
20. Korycki, A.; Garnier, C.; Bonmatin, M.; Laurent, E.; Chabert, F. Assembling of Carbon Fibre/PEEK Composites: Comparison of Ultrasonic, Induction, and Transmission Laser Welding. *Materials* **2022**, *15*, 6365. [CrossRef]
21. Li, W.; Palardy, G. Investigation of welding repair methods for thermoplastic composite joints. *Compos. Part B Eng.* **2023**, *264*, 110924. [CrossRef]
22. Wool, R.P.; O'Connor, K.M. A theory crack healing in polymers. *J. Appl. Phys.* **1981**, *52*, 5953–5963. [CrossRef]
23. Bastien, L.J.; Gillespie, J.W. A non-isothermal healing model for strength and toughness of fusion bonded joints of amorphous thermoplastics. *Polym. Eng. Sci.* **1991**, *31*, 1720–1730.
24. Grewell, D.; Benatar, A. Semiempirical, squeeze flow, and intermolecular diffusion model. II. Model verification using laser microwelding. *Polym. Eng. Sci.* **2008**, *48*, 1542–1549.
25. Acherjee, B.; Kuar, A.S.; Mitra, S.; Misra, D. Effect of carbon black on temperature field and weld profile during laser transmission welding of polymers: A FEM study. *Opt. Laser Technol.* **2012**, *44*, 514–521. [CrossRef]
26. de Gennes, P.G. Reptation of a Polymer Chain in the Presence of Fixed Obstacles. *J. Chem. Phys.* **1971**, *55*, 572–579. [CrossRef]
27. Doi, M.; Edwards, S.F. *The Theory of Polymer Dynamics*; International Series of Monographs on Physics; Oxford University Press: Oxford, NY, USA, 1988.
28. Szuchács, A.; Ageyeva, T.; Boros, R.; Kovács, J.G. Bonding strength calculation in multicomponent plastic processing technologies. *Mater. Manuf. Process.* **2022**, *37*, 151–159.
29. Wool, R.P. Chapter 8—Diffusion and autohesion. In *Adhesion Science and Engineering*; Dillard, D.A., Pocius, A.V., Chaudhury, M., Eds.; Elsevier Science B.V.: Amsterdam, The Netherlands, 2002; pp. 351–401. [CrossRef]
30. Giusti, R.; Lucchetta, G. Modeling the Adhesion Bonding Strength in Injection Overmolding of Polypropylene Parts. *Polymers* **2020**, *12*, 2063. [CrossRef] [PubMed]
31. Grewell, D.; Benatar, A. Semi-empirical, squeeze flow and intermolecular diffusion model. I. Determination of model parameters. *Polym. Eng. Sci.* **2008**, *48*, 860–867. [CrossRef]
32. Doi, M.; Edwards, S.F. Dynamics of concentrated polymer systems. Part 1.—Brownian motion in the equilibrium state. *J. Chem. Soc. Faraday Trans. 2 Mol. Chem. Phys.* **1978**, *74*, 1789–1801. [CrossRef]
33. Graessley, W.W. Some phenomenological consequences of the Doi–Edwards theory of viscoelasticity. *J. Polym. Sci. Polym. Phys. Ed.* **1980**, *18*, 27–34.
34. Wool, R.P. Adhesion at polymer–polymer interfaces: A rigidity percolation approach. *Comptes Rendus Chim.* **2006**, *9*, 25–44. [CrossRef]
35. Fetters, L.J.; Lohse, D.J.; Colby, R.H. Chain Dimensions and Entanglement Spacings. In *Physical Properties of Polymers Handbook*; Mark, J.E., Ed.; Springer: New York, NY, USA, 2007; pp. 447–454. [CrossRef]
36. Bex, G.J.P.; Ingenhut, B.L.J.; ten Cate, T.; Sezen, M.; Ozkoc, G. Sustainable approach to produce 3D-printed continuous carbon fiber composites: "A comparison of virgin and recycled PETG". *Polym. Compos.* **2021**, *42*, 4253–4264. [CrossRef]
37. Kim, J.K.; Han, C.D. Polymer-polymer interdiffusion during coextrusion. *Polym. Eng. Sci.* **1991**, *31*, 258–269.
38. Debondue, E.; Fournier, J.E.; Lacrampe, M.F.; Krawczak, P. Weld-line sensitivity of injected amorphous polymers. *J. Appl. Polym. Sci.* **2004**, *93*, 644–650. [CrossRef]
39. Lafranche, E.; Macedo, S.; Ferreira, P.; Martins, C.I. Thin wall injection-overmoulding of polyamide 6/polypropylene multilayer parts: PA6/PP-g-ma interfacial adhesion investigations. *J. Appl. Polym. Sci.* **2021**, *138*, 50294. [CrossRef]
40. Le, A.D.; Akué Asséko, A.C.; Nguyen, T.H.X.; Cosson, B. Laser intensity and surface distribution identification at weld interface during laser transmission welding of thermoplastic polymers: A combined numerical inverse method and experimental temperature measurement approach. *Polym. Eng. Sci.* **2023**. [CrossRef]
41. Akué Asséko, A.C.; Cosson, B.; Deleglise, M.; Schmidt, F.; Le Maoult, Y.; Lafranche, E. Analytical and numerical modeling of light scattering in composite transmission laser welding process. *Int. J. Mater. Form.* **2015**, *8*, 127–135. [CrossRef]
42. Liu, C.; He, J.; Ruymbeke, E.v.; Keunings, R.; Bailly, C. Evaluation of different methods for the determination of the plateau modulus and the entanglement molecular weight. *Polymer* **2006**, *47*, 4461–4479. [CrossRef]

Disclaimer/Publisher's Note: The statements, opinions and data contained in all publications are solely those of the individual author(s) and contributor(s) and not of MDPI and/or the editor(s). MDPI and/or the editor(s) disclaim responsibility for any injury to people or property resulting from any ideas, methods, instructions or products referred to in the content.

materials

MDPI

Article

A Study on Through-the-Thickness Heating in Continuous Ultrasonic Welding of Thermoplastic Composites

Bram C. P. Jongbloed *, Julie J. E. Teuwen, Rinze Benedictus and Irene Fernandez Villegas

Aerospace Structures and Materials Department, Faculty of Aerospace Engineering, Delft University of Technology, Kluyverweg 1, 2629 HS Delft, The Netherlands; J.J.E.Teuwen@tudelft.nl (J.J.E.T.); R.Benedictus@tudelft.nl (R.B.); I.FernandezVillegas@tudelft.nl (I.F.V.)
* Correspondence: B.C.P.Jongbloed@tudelft.nl

Abstract: Continuous ultrasonic welding is a promising technique for joining thermoplastic composites structures together. The aim of this study was to gain further insight into what causes higher through-the-thickness heating in continuous ultrasonic welding of thermoplastic composites as compared to the static process. Thermocouples were used to measure temperature evolutions at the welding interface and within the adherends. To understand the mechanisms causing the observed temperature behaviours, the results were compared to temperature measurements from an equivalent static welding process and to the predictions from a simplified heat transfer model. Despite the significantly higher temperatures measured at the welding interface for the continuous process, viscoelastic bulk heat generation and not thermal conduction from the interface was identified as the main cause of higher through-the-thickness heating in the top adherend. Interestingly the top adherend seemed to absorb most of the vibrational energy in the continuous process as opposed to a more balanced energy share between the top and bottom adherend in the static process. Finally, the higher temperatures at the welding interface in continuous ultrasonic welding were attributed to pre-heating of the energy director due to the vibrations being transmitted downstream of the sonotrode, to reduced squeeze-flow of energy director due to the larger adherend size, and to heat flux originating downstream as the welding process continues.

Keywords: fusion bonding; heat transfer; high-frequency welding; joining; CF/PPS; energy director

Citation: Jongbloed, B.C.P.; Teuwen, J.J.E.; Benedictus, R.; Villegas, I.F. A Study on Through-the-Thickness Heating in Continuous Ultrasonic Welding of Thermoplastic Composites. *Materials* **2021**, *14*, 6620. https://doi.org/10.3390/ma14216620

Academic Editor: Patricia Krawczak

Received: 25 September 2021
Accepted: 28 October 2021
Published: 3 November 2021

Publisher's Note: MDPI stays neutral with regard to jurisdictional claims in published maps and institutional affiliations.

Copyright: © 2021 by the authors. Licensee MDPI, Basel, Switzerland. This article is an open access article distributed under the terms and conditions of the Creative Commons Attribution (CC BY) license (https://creativecommons.org/licenses/by/4.0/).

1. Introduction

Because of their high specific mechanical properties, fibre-reinforced polymer composite materials are interesting for industries in which weight is of utmost importance, such as the aerospace industry. The polymers used as matrices in these composite materials are typically thermoset or thermoplastic. Thermoplastics, in contrast to thermosets, do not form a cross-linked network and they rely on the entanglement of linear polymer chains to form a solid structure. These thermoplastic polymer chains become mobile when heated sufficiently above the melting temperature for semi-crystalline polymers or above the glass transition temperature for amorphous polymers. This allows the use of cost-effective manufacturing and joining processes such as press forming and welding for thermoplastic composite parts and structures.

Welding has significant advantages over the more traditional joining methods mechanical fastening and adhesive bonding. For mechanical fastening drilling of holes results in breakage of fibres, and adhesive bonding requires intensive surface treatments and long curing cycles [1]. The most promising welding techniques for thermoplastic composite structures are resistance welding, induction welding, and ultrasonic welding [1–3]. The maturity levels of resistance and induction welding are the highest. Consequently, mainly these two technologies are currently being used in the aerospace industry, e.g., the rudder and elevator of the Gulfstream G650 [4] are assembled with induction welding, and the ribs and skin in the fixed leading edge of the A340 and A380 are joined using resistance welding [5]. Many recent studies have been published on process modelling and on experimental investigations to better understand the induction, resistance and ultrasonic welding processes [6–15], which illustrates the current relevance. Among the three mentioned welding techniques, ultrasonic welding is typically the fastest and lowest energy-consuming joining technique for thermoplastic composites [2]. During the ultrasonic welding process, a static force and high frequency/low amplitude mechanical vibrations are applied to the parts to be welded by means of a sonotrode (vibration phase of the process). Subsequently, the welded joint is allowed to cool down under pressure (consolidation phase). The mechanical vibrations promote the generation of heat through surface and viscoelastic friction [16–18]. Heat generation is focused at the weld interface by means of an energy director (ED) placed in between the adherends [17,19]. This ED consists of either one or more resin protrusions moulded on the surface of one of the adherends [20–22] or a loose resin-rich layer such as a film (also referred to as flat ED) [19,23] or a woven mesh [14,24] made from the same polymer material as that in the adherends. Due to the lower compressive stiffness of the ED compared to that of the fibre reinforced adherends, the ED undergoes higher cyclic strains and it therefore generates more heat than the adherends [22,25].

Ultrasonic welding of thermoplastic composites can be performed in either a static or a continuous fashion. Static ultrasonic welding, e.g., spot welding (Figure 1a) is defined as follows: a relatively small area is welded while both the welder and the adherends remain stationary during the vibration and consolidation phase. During the consolidation phase, the sonotrode itself applies a consolidation pressure. Sequential application of this static process allows obtaining multi-spot welded joints that can match load carrying capabilities of mechanically fastened joints [26,27]. The continuous ultrasonic welding process (Figure 1b) is defined as follows: a relatively large area is welded by continuously translating the welder with respect to the adherends or vice versa while exerting the ultrasonic vibrations and welding pressure [14,24]. Since the sonotrode continuously moves away from the just welded area, an additional consolidation shoe (consolidator) needs to be placed behind the sonotrode to allow the weld to cool down under pressure [15]. The continuous process has some benefits over the static process, i.e., a higher load can be transferred, and the joint is air and liquid tight, its maturity level is however lower. One of the main lessons learned so far in the development of continuous ultrasonic welding of thermoplastic composites is that the compliance of the energy director plays a more prominent role in ensuring weld uniformity than in the static process [24]. Likewise, heating and melting of the energy director extend beyond the footprint of the sonotrode in the continuous welding process and the temperatures at the welding interface are overall higher [14]. Finally, higher through-the-thickness or bulk heating occurs in continuous ultrasonic welding, especially in the adherend in direct contact with the sonotrode. As shown in Figure 2 [15], this leads to significant fibre and resin squeeze out, as well as accompanying porosity, upon application of the consolidation pressure.

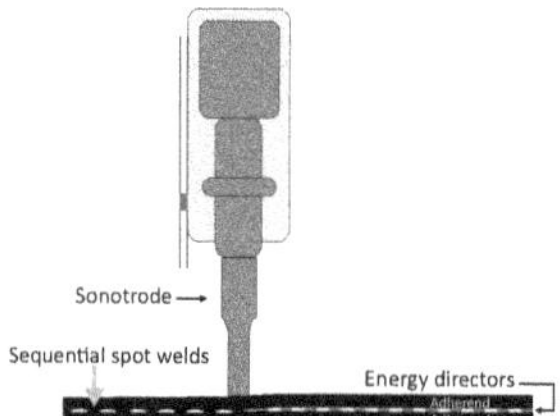

(a)

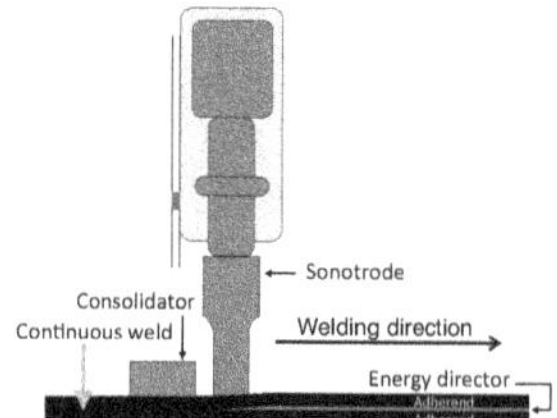

(b)

Figure 1. Schematics of static (**a**) and continuous (**b**) ultrasonic welding processes for thermoplastic composites.

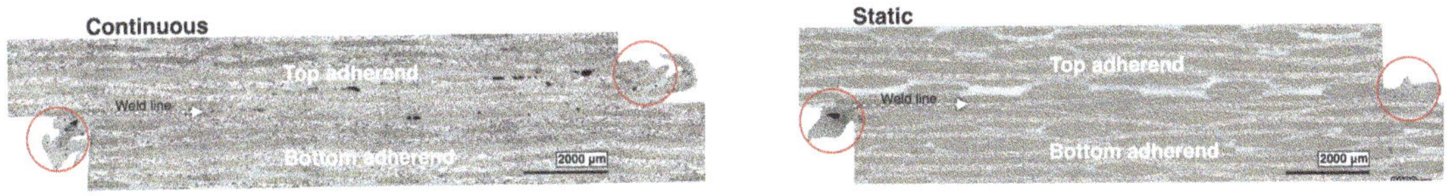

(a) (b)

Figure 2. Cross-sectional micrographs of a (**a**) continuous (adapted from [15]) and a (**b**) static ultrasonic weld. Welding force 500 N, vibration amplitude 80 µm, welding speed 35 mm/s (**a**), equivalent vibration time 430 ms (**b**) [14]. The same clamping jig and clamping configuration is used in both cases. The red circles indicate squeeze out of fibres and/or resin.

Excessive resin and fibre squeeze out are undesirable since they may compromise the structural integrity of the adherends and, hence, of the welded assembly. Consequently, the focus of the present study is gaining a better insight into what causes higher through-the-thickness heating in continuous ultrasonic welding of thermoplastic composites as compared to the static process. In addition, this leads to a better insight into the process and heating mechanisms and thus to high-quality welds. To this end, thermocouples were used to measure the temperature evolution within the adherends and at the weld interface for different welding configurations and combinations of welding parameters. The results were compared to temperature measurements from an equivalent static welding process and to the predictions from a simplified numerical heat transfer model. To further understand the through-the-thickness heating, the total energy input was also reduced by either increasing the welding speed or decreasing the vibration amplitude. Additionally, to understand heating ahead of the sonotrode a damping unit was introduced to dampen vibrations travelling downstream. Finally, the effect of the adherend size, i.e., large adherends for the continuous process versus small coupon sized adherends for the static process, on the temperature evolution was studied by means of static welds on different sized adherends.

2. Experimental Procedures

2.1. Materials

The thermoplastic composite laminates used for the welding experiments in this study were made out of carbon fibre fabric (five harness satin weave) impregnated with polyphenylene sulphide powder, CF/PPS semipreg (CF 0286 127 Tef4 43% from Toray Advanced Composites, Nijverdal, The Netherlands). The laminates were stacked according to a $[0/90]_{3S}$ sequence and subjected to a consolidation process in a hot platen press for 20 min at 320 °C and 1 MPa pressure, which, based on previous experience, results in void-free laminates. The consolidated laminates had a size of 580 mm by 580 mm and a thickness of approximately 1.85 mm. Different size adherends measuring 220 mm × 101.6 mm and 15 mm × 101.6 mm were cut from the consolidated laminates using a water jet cutter. For both adherend sizes, the main apparent fibre orientation was in the 101.6 mm direction. A 0.20 mm thick woven PPS mesh energy director with 37% open area (PPS100, supplied by PVF GmbH, Markt Schwaben, Germany) was used in all experiments to focus heat generation at the welding interface [14,24].

2.2. Continuous Ultrasonic Welding

The custom-built ultrasonic welding machine shown in Figure 3a was used for the continuous ultrasonic welding experiments. It consists of a stiff frame with an X-Y table on a guiding system, an off-the-shelf ultrasonic welder (VE20 SLIMLINE DIALOG 6200, Herrmann Ultrasonics, Karlsbad, Germany), and a custom-built consolidation unit. The welder vibrates at a fixed frequency of 20 kHz. Both the welder and consolidation unit are connected to the stiff frame. A rectangular sonotrode with a 15 mm × 27 mm contact area and a maximum peak-to-peak operational amplitude of 80 µm was used in the welding setup. The consolidation unit consisted of a 1.5 kN servo press kit (YJKP, Festo, Delft, The Netherlands), a stabilization guide unit to avoid sideways deflections, and a

40 mm × 30 mm copper block (Figure 3b). The consolidation force could be adjusted to a maximum of 1500 N. The distance between the consolidator and the sonotrode was set to 86.4 mm, based on the results of our previous study [15]. During the welding process, the adherends were translated with respect to the sonotrode generally following the welding direction I (Figure 3b), while the sonotrode continuously exerted the static welding force and vibrations, and the consolidator applied pressure. For some experiments the consolidation unit was used as a damping unit for which it was placed 18.4 mm ahead of the sonotrode (Figure 3b) and welding took place in the opposite direction (II). An aluminium base with bar clamps was used to clamp the adherends during the welding process (Figure 3b). Figure 3c shows the positioning of the bar clamps and of the sonotrode relative to the overlap. Note that this configuration, which we found to provide a more uniform temperature distribution across the overlap [15], differs slightly from the one used in [14]. The different sets of welding parameters used in this study are shown in Table 1.

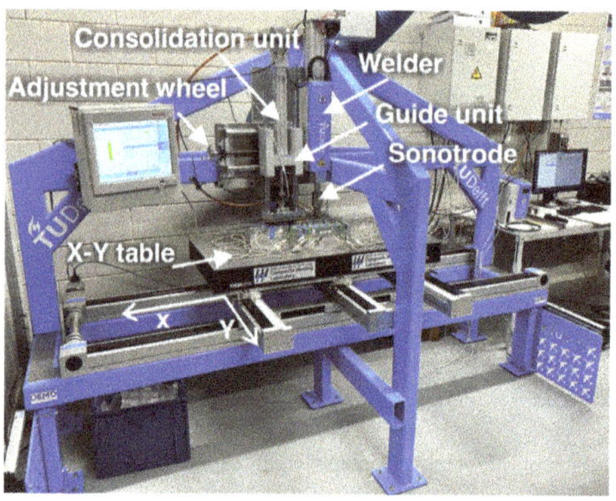

(a)

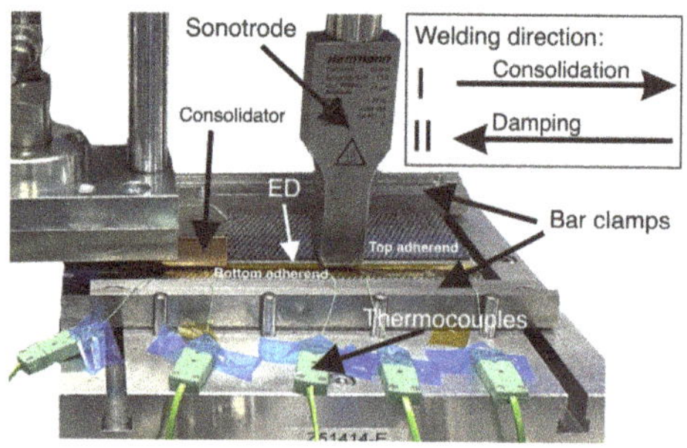

(b)

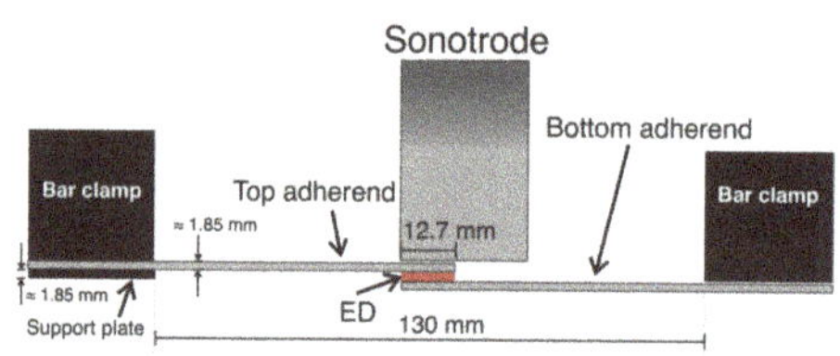

(c)

Figure 3. (**a**) Custom-built welding machine for static and continuous ultrasonic welding, (**b**) close-up of the continuous welding set-up visualising relative sonotrode and consolidator placement, and (**c**) schematic side-view of clamping distance and sonotrode placement.

2.3. Static Ultrasonic Welding

The same experimental setup as described in the previous subsection (Figure 3) was used for the static welding process. In this case, however, the adherends remained stationary and the consolidation unit was not used. The welding parameters used in the static process are shown in Table 1. The welder was set to vibrate for 430 ms, after which the vibrations stopped (end of vibrations), and the consolidation phase was initiated. As mentioned in the introduction, in the static process the sonotrode also provides the consolidation pressure during the consolidation phase. Note that the vibration time used for the static process, 430 ms, is equivalent to 35 mm/s welding speed in a continuous welding process with a 15 mm wide sonotrode. Static welds on the 15 mm wide adherends covered the entire overlap, whereas, in the case of 220 mm wide adherends, a single 15 mm wide static weld was created in the middle of the 220 mm wide overlap. During the static welding process, the vertical position of the sonotrode was monitored at a 1 kHz sampling frequency.

Table 1. Overview of Continuous (CUW) and Static (SUW) Ultrasonic Welding Experiments (Note: the amplitude (amp) values are peak-to-peak).

Process	Welding Parameters (Speed/Time, Force, Amp)	Consolidation Parameters	Number of Welds	Adherend Size (mm)	Config (Figure 4)	Remarks
CUW	35 mm/s, 500 N, 80 μm	800 N (1.6 MPa)	1	220 × 101.6	B	Setup with consolidation unit (Figure 3b). Reference case.
	35 mm/s, 500 N, 80 μm	800 N (1.6 MPa)	1	220 × 101.6	C	Setup with consolidation unit (Figure 3b).
	35 mm/s, 500 N, 80 μm	800 N (1.6 MPa)	3	220 × 101.6	D	Setup with consolidation unit (Figure 3b).
	35 mm/s, 500 N, 80 μm	800 N (1.6 MPa)	1	220 × 101.6	B	Setup with consolidation unit as damping unit (Figure 3b).
	65 mm/s, 500 N, 80 μm	800 N (1.6 MPa)	1	220 × 101.6	B	Setup with consolidation unit (Figure 3b). Higher speed case.
	35 mm/s, 500 N, 70 μm	800 N (1.6 MPa)	1	220 × 101.6	B	Setup with consolidation unit (Figure 3b). Lower amp case.
SUW	430 ms, 500 N, 80 μm	300 N (1.6 MPa) for 4 s	6	15 × 101.6	E	-
	430 ms, 500 N, 80 μm	300 N (1.6 MPa) for 4 s	3	15 × 101.6	F	-
	430 ms, 500 N, 80 μm	300 N (1.6 MPa) for 4 s	2	220 × 101.6	A	Static weld at TC location

2.4. Temperature Measurements

Temperatures were measured at the welding overlap and within the adherends using K-type thermocouples (GG220-2K-0, product number 2-2200-0004, Tempco B.V., Bodegraven, the Netherlands). The sleeved thermocouples had a total diameter of 0.70 mm, while the diameter of the thermocouple wires was 0.10 mm. An analogue thermocouple output amplifier (Adafruit AD8495) was used to simultaneously sample temperature readings at 1 kHz from a maximum of five thermocouples. A moving average filter (10 points for the static process and 25 or 60 points for the continuous process) was applied in MATLAB to filter out high-frequency fluctuations from the temperature data. For the temperature measurements at the weld interface, the thermocouples were placed in the middle of the overlap, sandwiched between the bottom adherend and the energy director. For the measurements within the adherends, the thermocouples were inserted into 0.7 mm diameter holes drilled up to a depth of approximately 7 mm. The placement and depth of each hole ensured that the tip of the thermocouple was located approximately in the middle of the overlap width (directly above or below the thermocouple at the weld interface) and approximately midway through the thickness of the adherend. Note that the diameter of the hole and of the sleeved thermocouple were the same to ensure a press fit. To measure the temperature at the weld interface and through the thickness in the top or bottom adherends, different thermocouple configurations shown in Figure 4 were used for the static (configurations A, E, and F) and continuous welding process (configurations B, C and D). Table 1 shows which of these thermocouple configurations were used in the different experiments.

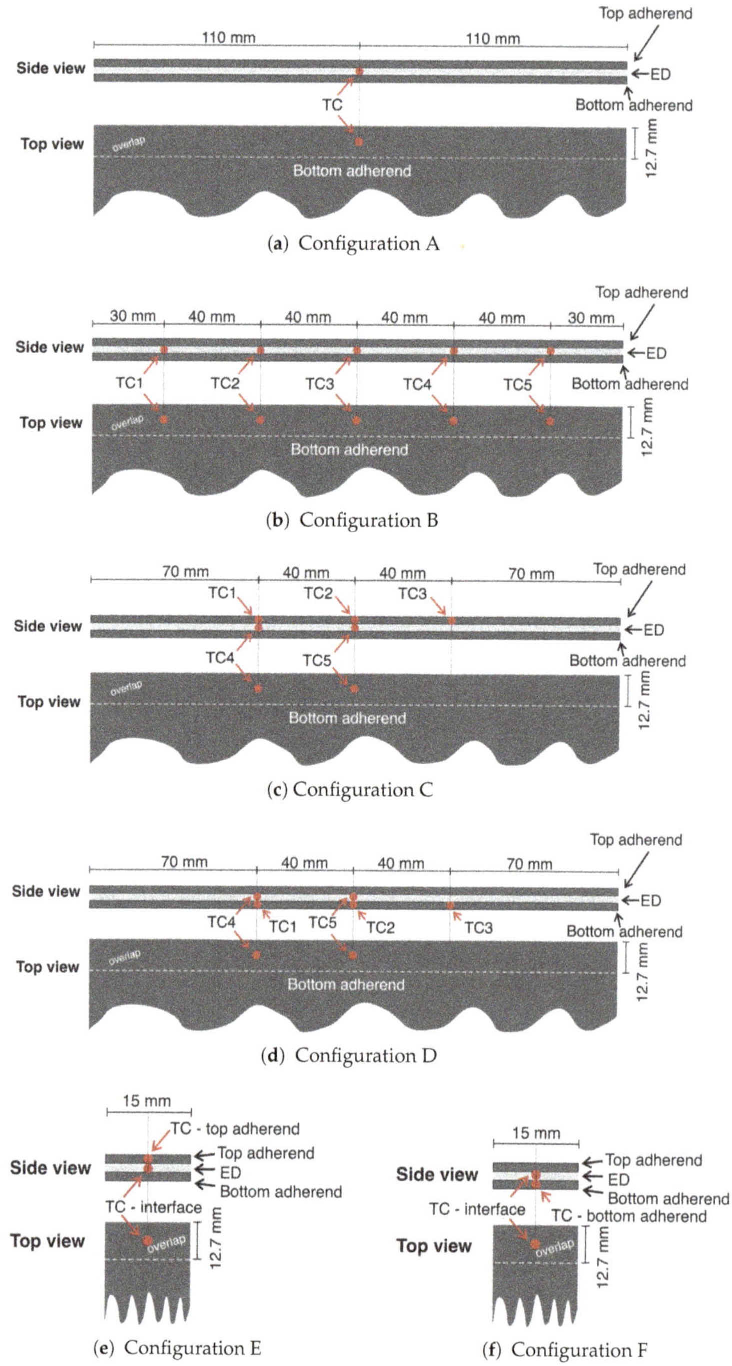

Figure 4. Schematic side and top view of the temperature measurement configurations used in this study on 220 mm wide and 15 mm wide adherends.

2.5. Testing and Analysis Techniques

Whenever necessary, the continuously welded adherends were cut into six 25.4 mm wide single lap shear samples with a diamond saw of which five were used for mechanical testing. The remaining sample was used for cross-sectional microscopy. The 28.8 mm wide edges at the start and at the end of the continuous welds were discarded. The single-lap shear samples were mechanically tested with a Zwick/Roell 250 kN (Kennesaw, GA, USA) universal testing machine with a cross-head speed of 1.3 mm/min. The grips were given the necessary offset to minimize secondary bending. The apparent lap shear strength (LSS) was calculated by dividing the maximum load by the overlap area. After mechanical testing, a non-contact roughness measurement/profiler system (Keyence VR-5000, Mechelen, Belgium) was used for the analysis of fracture surfaces. To obtain cross-sectional views from the welded adherends, specimens were cut and embedded in epoxy resin. They were ground and polished with a Struers Tegramin-20 polisher (Ballerup, Denmark). A 3D laser scanning confocal microscope (Keyence VK-X1000, Mechelen, Belgium) was used for inspecting the cross-sections.

2.6. Heat Transfer Model

A 2D transient heat transfer model representing the static welding process was created in COMSOL Multiphysics 5.5 (Burlington, MA, USA) to estimate the temperature increase in the adherends resulting from only heat conduction from the weld interface in both the static and the continuous process. Since the energy director is very thin compared to the adherends, it was not included in the model. The heat transfer model is based on the following heat transfer equation:

$$\rho C_p \frac{\partial T}{\partial t} = k \nabla^2 T \quad (1)$$

in which T is temperature, t is time, and ρ, C_p, and k are the density, thermal capacity, and thermal conductivity of CF/PPS, respectively.

Figure 5 shows the geometry and boundary conditions and Table 2 lists the material properties used in the model as per Equation (1). The outer boundaries were thermally insulated and the initial temperature of the sonotrode and base was 20 °C. The experimental temperature data obtained from measurements at the welding interface was applied as an input to the entire welding interface in the model. The resulting temperature evolution predicted by the model was evaluated midway through the thickness of the top and of the bottom adherends. It should be noted that due to the conservative nature of the boundary conditions (i.e., no heat dissipation to the environment) as well as the room-temperature material properties used for the CF/PPS material, the model most likely overestimates the heat transferred from the weld interface to the adherends and hence higher overall temperatures when used for the static process. Nevertheless, by comparing the temperature evolution provided by the model and the experimental value, the simplified model is still a valuable tool to discern whether the cause of through-the-thickness heating in the adherends is heat transfer or there are other sources. When used for the continuous process, however, the following simplifications in the model: (i) the sonotrode is in contact with the top adherend during the entire cooling phase, and (ii) there is no heat flux from downstream and upstream of the sonotrode into the studied area, can be expected to somewhat offset this overestimation. It should be noted that the model is not suitable for accurately predicting the actual temperature values, since that would require a 3D model with temperature-dependent material properties and in case of the continuous process a more complicated dynamic model with a moving welder head.

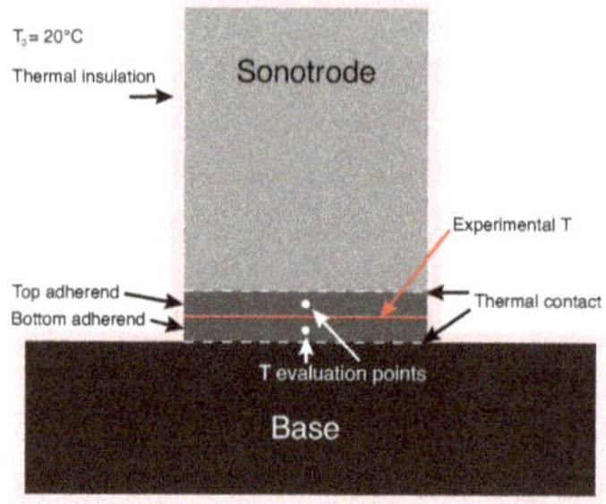

Figure 5. Schematic model with boundary conditions for heat transfer model of static welding setup.

Table 2. Material Properties used for Model.

	CF/PPS Adherends	
Property	**Value**	**Remarks**
Density (ρ)	1540 kg/m^3	Calculated based on weight measurements
Heat capacity (C_p)	681 J/(kg·°C)	Value taken at 20 °C [28]
Thermal conductivity (k)	0.34 W/(m·°C)	Value taken at 20 °C [28]
	Aluminum Base	
Property	**Value**	**Remarks**
Density (ρ)	2665 kg/m^3	From COMSOL material library of aluminum 5083
Heat capacity (C_p)	955 J/(kg·°C)	From COMSOL material library of aluminum 5083 Value shown at 20 °C
Thermal conductivity (k)	120 W/(m·°C)	From COMSOL material library of aluminum 5083 Value shown at 20 °C
	Steel Sonotrode	
Property	**Value**	**Remarks**
Density (ρ)	7860 kg/m^3	From COMSOL material library of steel 1040
Heat capacity (C_p)	480 J/(kg·°C)	From COMSOL material library of steel 1040 Value shown at 20 °C
Thermal conductivity (k)	52 W/(m·°C)	From COMSOL material library of steel 1040 Value shown at 20 °C

3. Results

The temperature evolution at the weld interface for the continuous process is shown in Figure 6 for the reference case, higher welding speed case, lower amplitude case, and consolidator as damping case (Figure 6a, 6b, 6c and 6d, respectively). Figure 7 shows temperature and vertical sonotrode displacement curves for static welds on (a) 15 mm wide and (b) 220 mm wide adherends. Figures 8 and 9 show the temperature evolution at the weld interface and through the thickness for the (a) top and (b) bottom adherends in the continuous and static process, respectively. It should be noted that some thermocouples malfunctioned during the continuous process (Figure 4c,d): TC2 for all welds and multiple thermocouples in one weld for configuration D (Table 1). Therefore, these results have been omitted. The measurements within the top adherend (Figure 8a) might be less trustworthy as severe overheating was observed of this adherend at the thermocouple locations and thermocouples TC3 and TC4 partially malfunctioned (as seen by the temperature spikes in Figure 8a). Figure 10 shows the modelled temperature prediction due to heat transfer to the middle of the adherends for the static welding setup (indicated in Figure 5) based on the experimental temperature input at the weld interface from (a) the continuous and (b) the static welding process. It should be noted that for the top adherend in the continuous process no experimental value is shown because, as mentioned before, the top adherend was severely overheated at the thermocouple locations making the results less trustworthy. Figure 11 shows cross-sectional micrographs from continuous ultrasonic welds for (a) the higher speed case and (b) the lower amplitude case. Note that for the reference case the cross-section is shown in Figure 2a. Representative fracture surfaces from the reference, higher speed, and lower amplitude cases are shown in Figure 12a, 12b, and 12c, respectively. The single-lap shear strength values for the continuous and static welds are shown in Table 3.

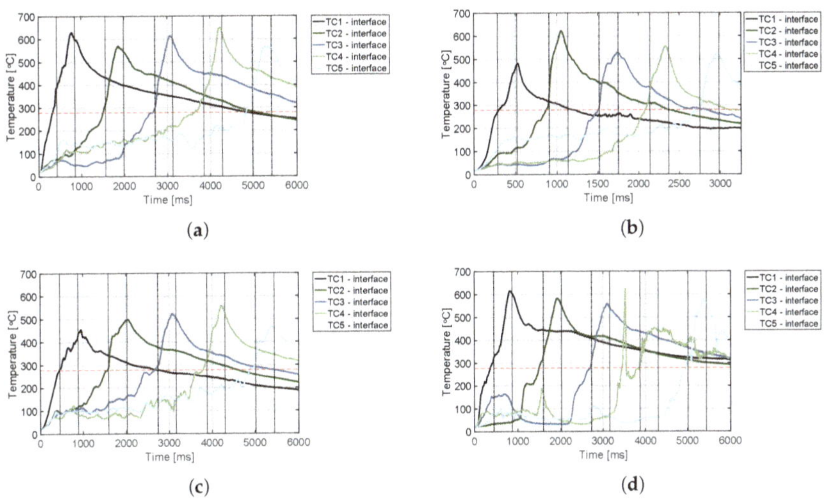

Figure 6. Temperature evolution for the continuous welding process at the weld interface for: (**a**) reference case, (**b**) higher welding speed case (65 mm/s), (**c**) lower amplitude case (70 μm), and (**d**) damping case. TC1 to TC5 were respectively located under the sonotrode during the five grey areas. The red dashed line indicates the melting temperature of PPS (T_m, 280 °C) as experimentally determined by DSC analysis.

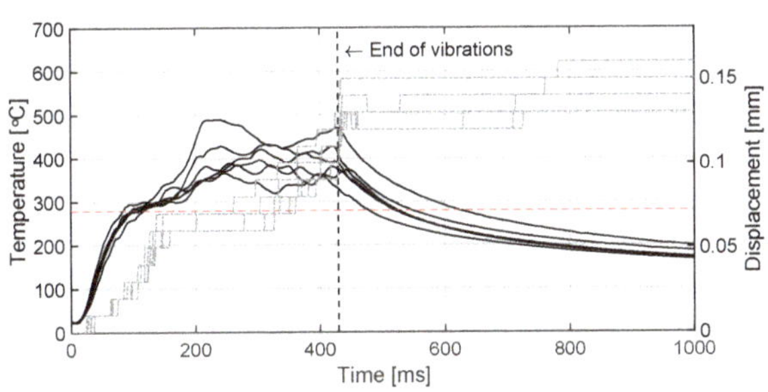

(**a**)

Figure 7. *Cont.*

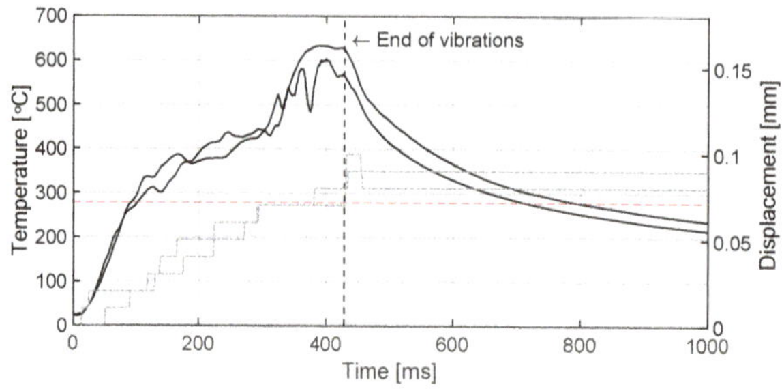

(**b**)

Figure 7. Interface temperature (black curves) and measured downward vertical displacement (grey curves) of the sonotrode for static welds on (**a**) 15 mm wide and (**b**) 220 mm wide adherends. It should be noted that one of the two welds in (**b**) was made 40 mm to the left from the intended location in Figure 4a. The red dashed line indicates the melting temperature of PPS (T_m, 280 °C) as experimentally determined by DSC analysis.

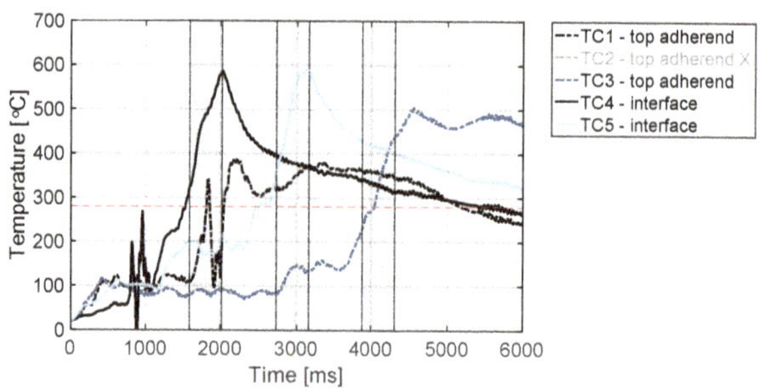

(**a**)

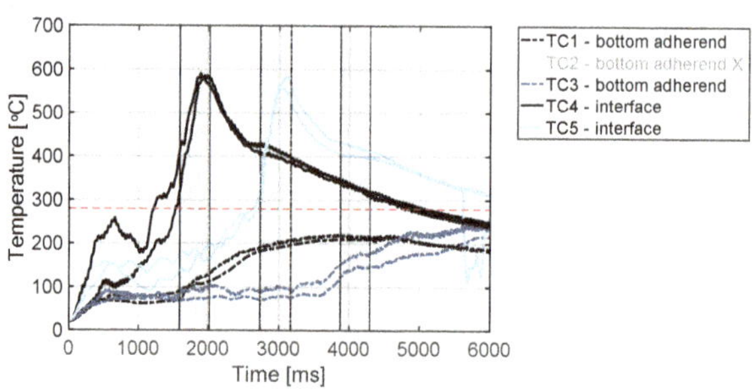

(**b**)

Figure 8. Temperature evolution for the continuous welding process at the weld interface and through the thickness for (**a**) the top adherend and (**b**) the bottom adherend. The grey areas indicate the time span during which a specific thermocouple was located under the sonotrode. The red dashed line indicates the melting temperature of PPS (T_m, 280 °C) as experimentally determined by DSC analysis.

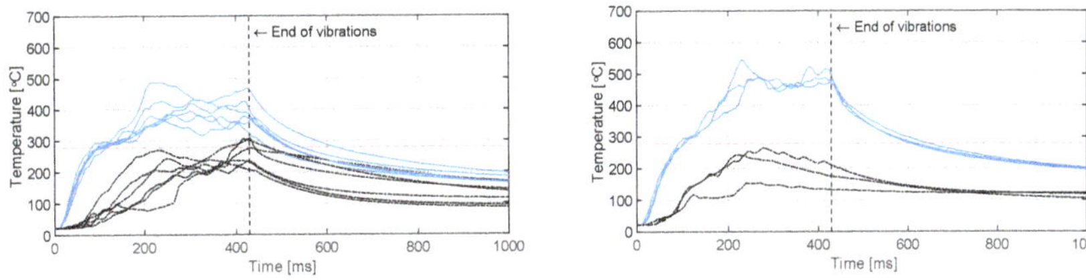

(a) (b)

Figure 9. Temperature evolution for the static welding process at the weld interface (blue curves) and through the thickness (black curves) for (**a**) the top adherend and (**b**) the bottom adherend. The temperature evolution at the weld interface in (**a**) is the same as Figure 7a. The red dashed line indicates the melting temperature of PPS (T_m, 280 °C) as experimentally determined by DSC analysis.

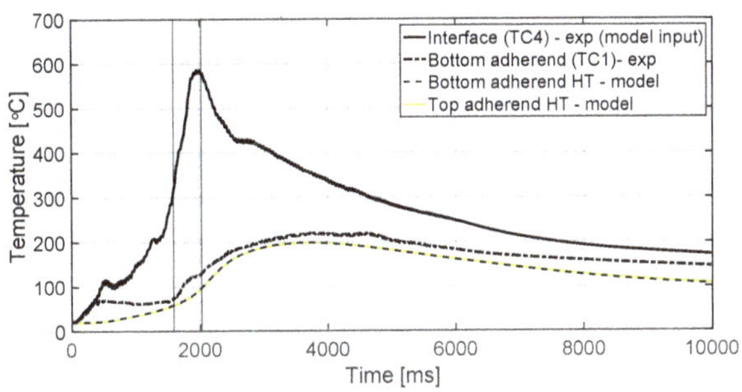

(a)

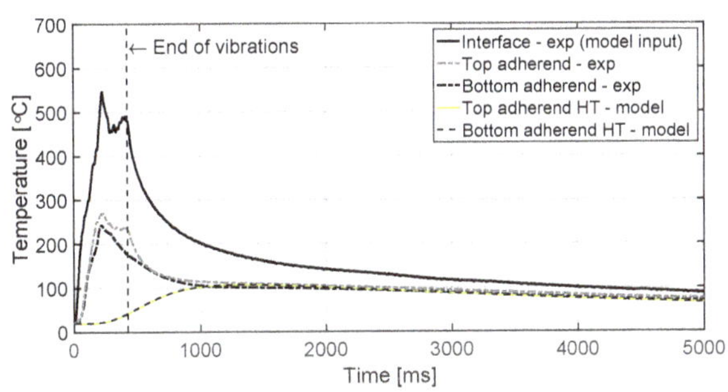

(b)

Figure 10. Modelled through-the-thickness temperature evolution due to heat transfer (HT) as a result from experimental (exp) temperature input at the weld interface for (**a**) the continuous and (**b**) static temperature evolution together with a representative experimental temperature evolution as a reference. The grey area (**a**) and the vertical line (**b**) indicate the time span during which the thermocouples were located under the sonotrode. Note that in (**a**) no experimental temperature evolution for the top adherend is shown as it was deemed less trustworthy due to severe overheating.

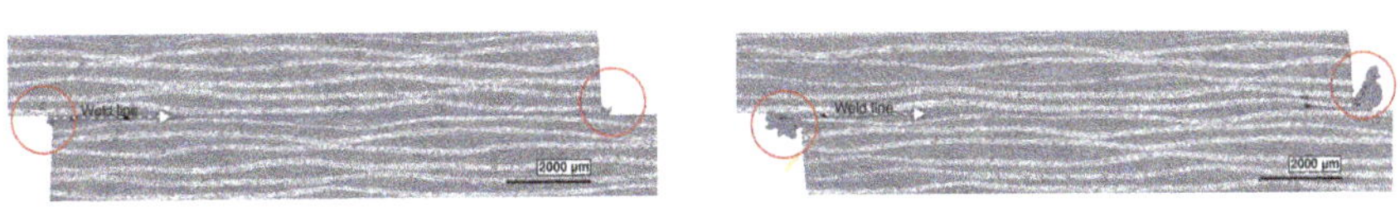

(a) (b)

Figure 11. Representative cross-sectional micrographs of continuous ultrasonic welds for (**a**) higher speed case (65 mm/s), and (**b**) lower amplitude case (70 μm). Red circles indicate squeeze-out location.

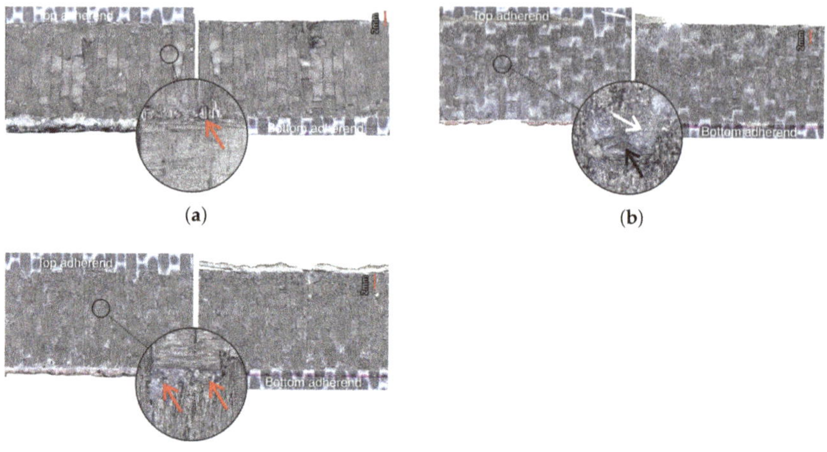

Figure 12. Representative fracture surfaces from continuous welds for (a) the reference case, (b) higher welding speed case (65 mm/s), and (c) the lower amplitude case. Red arrows indicate voids, white arrow indicates unwelded area together with voids, and black arrow indicates area without connection between top and bottom adherends.

Table 3. Average Lap Shear Strength (LSS) Values with Standard Deviation for Different Continuous and Static Welds.

Process	Parameters (Speed/Time, Force, Amplitude)	LSS (MPa)	Remarks
CUW	35 mm/s, 500 N, 80 μm	39.6 ± 2.3 (n = 5)	Reference case. Samples tested from configuration B with consolidation unit (Figure 3b).
CUW	35 mm/s, 500 N, 70 μm	37.2 ± 2.5 (n = 5)	Lower amplitude case. Samples tested from configuration B with consolidation unit (Figure 3b).
CUW	65 mm/s, 500 N, 80 μm	25.0 ± 3.9 (n = 5)	Higher speed case. Samples tested from configuration B with consolidation unit (Figure 3b).
SUW	440 ms, 500 N, 80 μm	34.3 ± 1.2 (n = 4)	Value taken from [14].

4. Discussion

The aim of this study was gaining further insight into what causes higher through-the-thickness heating in continuous than in static ultrasonic welding. Through-the-thickness heating combined with the application of the welding/consolidation pressure may cause matrix and fibre squeeze flow as well as porosity in the adherends [15] and hence may affect their structural integrity. Firstly, the heating mechanisms responsible for through-the-thickness heating were investigated by comparing temperature measurements from the static and continuous process to heat transfer model predictions. Secondly, it was studied what causes the temperature differences between the static and continuous process. Finally, the quality of the welds was discussed in view of the temperature evolution.

A plausible cause of higher through-the-thickness heating is the higher overall temperatures at the welding interface (Figure 13) resulting in increased thermal conduction to the adherends. The predictions of the heat transfer model (Figure 10) show indeed that the higher interface temperatures in the continuous process do result in higher temperature increase caused by conduction in the middle of the adherends. However, there are notorious differences between measured and predicted temperature evolutions for the adherends in the static process (Figure 10b) and presumably for the top adherend in the continuous process (Figures 8a and 10a). These differences indicate the existence of an extra heating mechanism responsible for the steep temperature increase of the adherends during the heating phase of the welding process (Figures 8a and 9) in addition to the relatively gentle temperature increase caused by thermal conduction observed in the model predictions (Figure 10), which peaks during the cooling phase. This extra mechanism is bulk viscoelastic heating and its prevalent role in the temperature evolution in the adherends is not surprising since, once compressed, the energy director is very thin (≈0.10 mm) and hence it has a low compliance [24] and low ability to concentrate viscoelastic heat generation at the interface only [23].

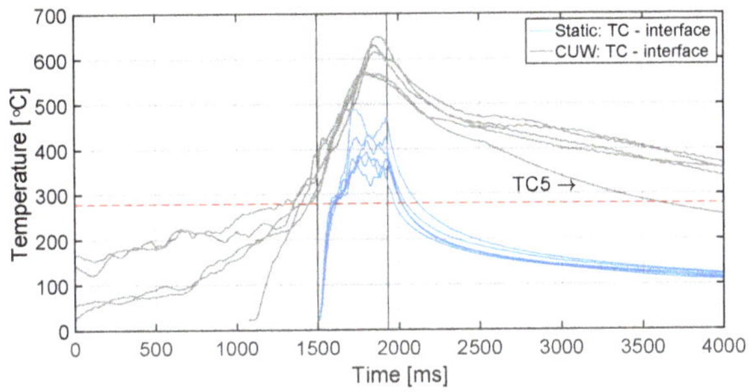

Figure 13. Combined interface temperature evolutions of continuous (reference case) (Figure 6a) and static (Figure 9a) processes superimposed for time that thermocouple experiences vibrations directly under sonotrode. The red dashed line indicates the melting temperature of PPS as experimentally determined by DSC analysis.

Contrarily, temperature measurements in the bottom adherend during the continuous process (Figure 8b) show trends that are closer to what one could expect from thermal conduction from the interface (Figure 10a). This observation, together with the overheating observed at the locations of the thermocouples within the top adherend (Figure 8a), prompts us to think that in the continuous process the top adherend absorbs significantly more vibration energy than the bottom adherend. This is contrary to a more balanced energy distribution in the static process (Figure 9). This is consistent with the top adherend experiencing more severe heating than the bottom adherend and than any of the adherends in the static process under equivalent process parameters (Figure 2). Reducing the total vibration energy associated with the process, by either increasing the welding speed or decreasing the vibration amplitude, has indeed a substantial effect on reducing through-the-thickness heating and the corresponding matrix and fibre squeeze flow in the top adherend (Figure 11) supporting our hypothesis. The cause of the uneven distribution of vibration energy between the top and bottom adherend in the continuous process is yet unknown. We believe that the translation of the vibrating sonotrode on the surface of the top adherend might cause an extra component of cyclic strains parallel to the welding interface resulting in a superposition of viscoelastic heating sources in that adherend. Further research should however be performed to test this hypothesis.

Additionally, given the importance of the temperature evolution at the welding interface in the quality of the welded joints, it is important to understand what causes the temperature differences between the continuous and the static process (Figure 13) and their relevance in the outcome of the welding process. It should be noted that the temperature differences between the two processes shown in Figure 13 are much higher than those reported in our previous work [14]. We believe such seeming inconsistency stems from the different clamping schemes in both studies; in particular differences in the clamping distance, which is known to affect the cycling strains and, hence, heat generation in the energy director [29].

Firstly, the pre-heating experienced by the energy director, which is consistent with previous observation of thermal effects in the energy director downstream of the sonotrode [14], is believed to result from the ultrasonic vibration being transmitted downstream of the sonotrode. Far from the sonotrode, surface friction between thermocouples, energy director and adherends most likely causes the temperature to almost instantaneously increase to around 100 °C at the onset of the welding process. Once the sonotrode is close enough for the stiff adherends to exert sufficient pressure and, hence, sufficient cyclic strain in the energy director, viscoelastic heating accounts for a faster temperature increase. Indeed damping of the vibrations beyond the sonotrode effectively minimizes pre-heating far from the sonotrode (Figure 14). It does not, however, remove pre-heating as the sonotrode gets closer to the measuring location owing to limitations in the experimental setup (i.e., the minimum practical distance between the damping unit and the sonotrode).

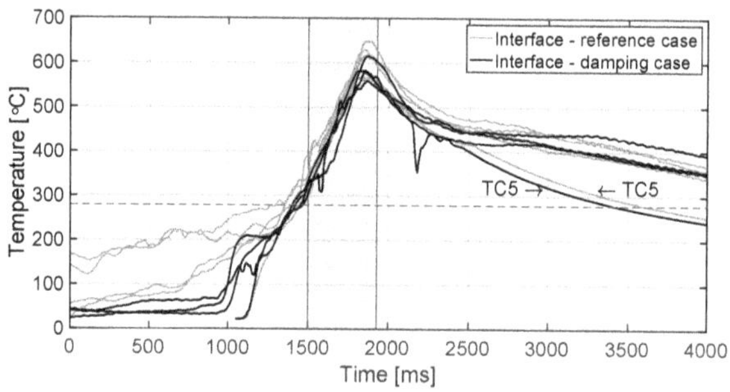

Figure 14. Combined interface temperature evolutions of the continuous ultrasonic welding process for reference case (Figure 6a) and damping case with the consolidator as damping unit (Figure 6d without TC4) superimposed for time that thermocouple experiences vibrations directly under sonotrode (grey area). The red dashed line indicates the melting temperature of PPS as experimentally determined by DSC analysis.

Secondly, the temperature at the interface experiences a continuous increase during the time the sonotrode is moving above a certain location which results in higher maximum temperatures as compared to the static process (Figure 13). Based on the results of the static welds on the same adherend size used for the continuous process (Figure 7), this continuous temperature increase can be attributed to the limitations imposed by the adherends to the squeeze flow of the energy director (as seen in a lower total displacement) and, consequently, limitations to the associated cooling effect (as seen in the continuous temperature increase) [23].

Thirdly, longer times needed to cool down the interface below T_m are associated with the heat flux originated downstream as the welding process progresses, as indicated by the consistently faster cooling measured by the last thermocouple (TC5) in the weld line (i.e., the one with the lowest influence by downstream heating, Figures 13 and 14). It is worth noticing that before this effect becomes relevant (i.e., beyond divergence point between thermocouple measurements in Figures 13 and 14), there is an initial faster cooling stage most likely influenced by the temperature in the composite layers in close proximity to the interface. In fact, the process that results in the lowest matrix and fibre squeeze out from the top adherend, i.e., higher speed case, is the one showing a significantly higher cooling rate in that initial stage (Figure 15a).

Finally, regarding the quality of the welds, it is interesting to note that in those cases in which the strength of continuously welded joints is comparable (or higher) to that of the static welds (i.e., reference and lower amplitude case, see Table 3), the average area delimited by the temperature curves and the melting temperature of PPS (presumably related to the actual thermal energy invested in the creation of the welded joints) is much higher than that in the static process (Figures 13 and 15b). Interestingly, in the case in which both areas are closer to each other (i.e., higher speed case, Figure 15a) the strength of continuous joints is significantly lower in accordance with the presence of unwelded areas on the fracture surfaces (Figure 12b). These observations, which indicate that in the continuous process the thermal energy input required to create a weld is higher than in the static process, relate to the lower temperature in the bottom adherend (as discussed previously) hindering the creation of the welded joint.

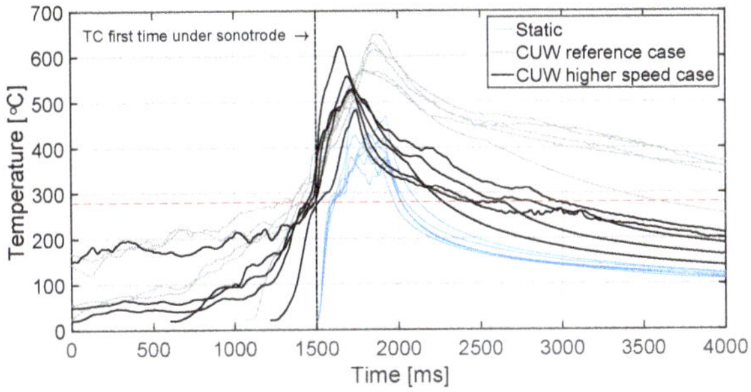

(a)

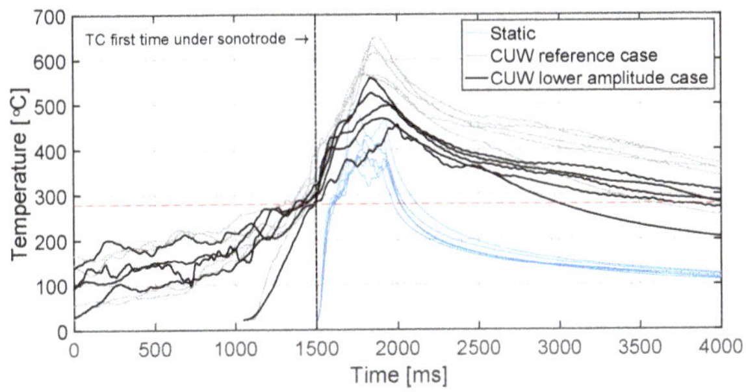

(b)

Figure 15. Combined interface temperature evolutions of the reference case (Figure 6a) of the continuous process, the static process (Figure 9a), and the temperature evolutions for (**a**) the higher welding speed case (65 mm/s) (Figure 6b), and (**b**) the lower amplitude case (70 μm) (Figure 6c). The temperature evolutions were superimposed at the moment they were under the sonotrode for the first time. The red dashed line indicates the melting temperature of PPS as experimentally determined by DSC analysis.

5. Conclusions

The aim of this study was to gain further insight into what causes higher through-the-thickness heating in continuous ultrasonic welding of thermoplastic composites as compared to the static process. In the continuous process, temperatures at the welding interface were indeed found to be significantly higher than in the static process. This was attributed to a combination of pre-heating of the energy director due to the vibrations being transmitted downstream of the sonotrode, reduced squeeze-flow of the energy director due to the larger adherend size, and heat flux originating downstream as the welding process continues. Thermal conduction from the hotter interface was however found to not be the main cause of higher through-the-thickness heating in the top adherend, which was in turn attributed to viscoelastic bulk heat generation. The top adherend seemed to indeed absorb most of the vibrational energy in the continuous process as opposed to a more balanced energy share between the top and bottom adherend in the static process. On the contrary, the bottom adherend showed a temperature evolution similar to what could be expected from predominant thermal conduction from the welding interface. Consequently, reducing the total vibration energy introduced in the material in the continuous welding process proved to have a substantial effect on reducing through-the-thickness heating.

Author Contributions: Conceptualization, B.C.P.J., J.J.E.T. and I.F.V.; methodology, B.C.P.J., J.J.E.T and I.F.V.; formal analysis, B.C.P.J.; investigation, B.C.P.J.; resources, R.B. and I.F.V.; writing—original draft preparation, B.C.P.J.; writing—review and editing, B.C.P.J., J.J.E.T. and I.F.V.; visualization B.C.P.J.; supervision, J.J.E.T., R.B. and I.F.V.; project administration, I.F.V.; funding acquisition, I.F.V. All authors have read and agreed to the published version of the manuscript.

Funding: This study was funded by the European research programme Clean Sky. The ecoTECH project has received funding from the European Union's Horizon 2020 Clean Sky 2 Joint Undertaking under the AIRFRAME ITD grant agreement number GAM AIR 2020—945521.

Institutional Review Board Statement: Not applicable.

Informed Consent Statement: Not applicable.

Data Availability Statement: The data presented in this study are available on request from the corresponding author.

Conflicts of Interest: The authors declare no conflict of interest.

References

1. Ageorges, C.; Ye, L.; Hou, M. Advances in fusion bonding techniques for joining thermoplastic matrix composites: A review. *Compos.-Part A Appl. Sci. Manuf.* **2001**, *32*, 839–857. [CrossRef]
2. Villegas, I.F.; Moser, L.; Yousefpour, A.; Mitschang, P.; Bersee, H.E. Process and performance evaluation of ultrasonic, induction and resistance welding of advanced thermoplastic composites. *J. Thermoplast. Compos. Mater.* **2013**, *26*, 1007–1024. [CrossRef]
3. Yousefpour, A.; Hojjati, M.; Immarigeon, J. Fusion Bonding/Welding of Thermoplastic Composites. *J. Thermoplast. Compos. Mater.* **2004**, *17*, 303–341. [CrossRef]
4. Offringa, A. New thermoplastic composite design concepts and their automated manufacture. *JEC Compos. Mag.* **2010**, *58*, 45–49.
5. Gardiner, G. Thermoplastic composites gain leading edge on the A380. *High-Perform. Compos.* **2006**, *14*, 50–55.
6. Grouve, W.; Vruggink, E.; Sacchetti, F.; Akkerman, R. Induction heating of UD C/PEKK cross-ply laminates. *Procedia Manuf.* **2020**, *47*, 29–35. [CrossRef]
7. Lionetto, F.; Pappadà, S.; Buccoliero, G.; Maffezzoli, A. Finite element modeling of continuous induction welding of thermoplastic matrix composites. *Mater. Des.* **2017**, *120*, 212–221. [CrossRef]
8. Brassard, D.; Dubé, M.; Tavares, J.R. Resistance welding of thermoplastic composites with a nanocomposite heating element. *Compos. Part B Eng.* **2019**, *165*, 779–784. [CrossRef]
9. Lundström, F.; Frogner, K.; Andersson, M. Analysis of the temperature distribution in weave-based CFRP during induction heating using a simplified numerical model with a cross-ply representation. *Compos. Part B Eng.* **2021**, *223*, 109153. [CrossRef]
10. Barazanchy, D.; van Tooren, M. Heating mechanisms in induction welding of thermoplastic composites. *J. Thermoplast. Compos. Mater.* **2021**, 08927057211011621,
11. Nagel, L.; Herwig, A.; Schmidt, C.; Horst, P. Numerical Investigation of Residual Stresses in Welded Thermoplastic CFRP Structures. *J. Compos. Sci.* **2021**, *5*, 45, [CrossRef]
12. Xiong, X.; Wang, D.; Ren, R.; Zhao, P.; Cui, X.; Ji, S. Improved mechanical properties of resistance welded joints of thermoplastic composites by versatile graphene oxide. *J. Adhes. Sci. Technol.* **2021**, *35*, 1099–1113. [CrossRef]
13. Köhler, F.; Villegas, I.; Dransfeld, C.; Herrmann, A. Static ultrasonic welding of carbon fibre unidirectional thermoplastic materials and the influence of heat generation and heat transfer. *J. Compos. Mater.* **2021**, *55*, 2087–2102. [CrossRef]
14. Jongbloed, B.; Teuwen, J.; Benedictus, R.; Villegas, I.F. On differences and similarities between static and continuous ultrasonic welding of thermoplastic composites. *Compos. Part B Eng.* **2020**, *203*, 108466. [CrossRef]
15. Jongbloed, B.C.P.; Vinod, R.; Teuwen, J.J.E.; Benedictus, R.; Villegas, I.F. (De)consolidation in Ultrasonic Welding of Thermoplastic Composites. *Compos. Part A Appl. Sci. Manuf.* **2021**, under peer review.
16. Zhang, Z.; Wang, X.; Luo, Y.; Zhang, Z.; Wang, L. Study on Heating Process of Ultrasonic Welding for Thermoplastics. *J. Thermoplast. Compos. Mater.* **2010**, *23*, 647–664. [CrossRef]
17. Benatar, A.; Gutowski, T. Ultrasonic welding of PEEK graphite APC-2 composites. *Polym. Eng. Sci.* **1989**, *29*, 1705–1721. [CrossRef]
18. Tolunay, M.; Dawson, P.; Wang, K. Heating and bonding mechanisms in ultrasonic welding of thermoplastics. *Polym. Eng. Sci.* **1983**, *23*, 726–733. [CrossRef]
19. Villegas, I. In situ monitoring of ultrasonic welding of thermoplastic composites through power and displacement data. *J. Thermoplast. Compos. Mater.* **2015**, *28*, 66–85. [CrossRef]
20. Villegas, I.; Valle Grande, B.; Bersee, H.; Benedictus, R. A comparative evaluation between flat and traditional energy directors for ultrasonic welding of CF/PPS thermoplastic composites. *Compos. Interfaces* **2015**, *22*, 717–729. [CrossRef]
21. Villegas, I.; Palardy, G. Ultrasonic welding of CF/PPS composites with integrated triangular energy directors: Melting, flow and weld strength development. *Compos. Interfaces* **2017**, *24*, 515–528. [CrossRef]

22. Benatar, A.; Eswaran, R.V.; Nayar, S.K. Ultrasonic welding of thermoplastics in the near-field. *Polym. Eng. Sci.* **1989**, *29*, 1689–1698. [CrossRef]
23. Palardy, G.; Villegas, I. On the effect of flat energy directors thickness on heat generation during ultrasonic welding of thermoplastic composites. *Compos. Interfaces* **2017**, *24*, 203–214. [CrossRef]
24. Jongbloed, B.; Teuwen, J.; Palardy, G.; Villegas, I.F.; Benedictus, R. Continuous ultrasonic welding of thermoplastic composites: Enhancing the weld uniformity by changing the energy director. *J. Compos. Mater.* **2019**. [CrossRef]
25. Levy, A.; Le Corre, S.; Villegas, I. Modeling of the heating phenomena in ultrasonic welding of thermoplastic composites with flat energy directors. *J. Mater. Process. Technol.* **2014**, *214*, 1361–1371. [CrossRef]
26. Zhao, T.; Broek, C.; Palardy, G.; Villegas, I.; Benedictus, R. Towards robust sequential ultrasonic spot welding of thermoplastic composites: Welding process control strategy for consistent weld quality. *Compos. Part A Appl. Sci. Manuf.* **2018**, *109*, 355–367. [CrossRef]
27. Zhao, T.; Rans, C.; Fernandez Villegas, I.; Benedictus, R. On sequential ultrasonic spot welding as an alternative to mechanical fastening in thermoplastic composite assemblies: A study on single-column multi-row single-lap shear joints. *Compos. Part A Appl. Sci. Manuf.* **2019**, *120*, 1–11. [CrossRef]
28. Villegas, I.F.; Bramon, J.V. *World Class Composite Solutions, TUDelft Work Package 3 Technical Report*; Aerospace Structures and Materials, Faculty of Aerospace Engineering, Delft University of Technology: Delft, The Netherlands, 2018.
29. Brito, C.B.G.; Teuwen, J.; Dransfeld, C.A.; Villegas, I.F. The effects of misaligned adherends on static ultrasonic welding of thermoplastic composites. *Compos. Part A Appl. Sci. Manuf.* **2021**, under peer review.

materials

MDPI

Article

Effect of Adherend Thickness on Near-Field Ultrasonic Welding of Single-Lap CF/LMPAEK Thermoplastic Composite Joints

Natalia Sofia Guevara-Sotelo and Irene Fernandez Villegas *

Aerospace Structures and Materials Department, Faculty of Aerospace Engineering, Delft University of Technology, 2629 HS Delft, The Netherlands; n.s.guevarasotelo@tudelft.nl
* Correspondence: i.fernandezvillegas@tudelft.nl

Abstract: Ultrasonic welding is a fast and promising joining technique for thermoplastic composite parts. Understanding how changing the part thickness affects the process is crucial to its future upscaling and industrialization. This article presents an initial insight into the effect of the adherend's thickness on the near-field ultrasonic welding of CF/LMPAEK thermoplastic composites. Different thicknesses of the top and bottom adherend were welded and analyzed using the output data of the welding equipment, temperature measurements, and other visual characterization techniques. Increasing the thickness of both the top and the bottom adherends showed to increase the power consumed during welding. An overshoot in the power needed at the onset of the welding process for increased thickness of the top adherend precluded welding beyond a threshold thickness of 4.72 mm. In the case of the thicker top adherends, there was also melting of the energy director and early fiber squeeze-out within the top adherend as a result of increased bulk heating. Increased bulk heating was hypothesized to be caused by increased hammering, as indicated by the amplitude readings for thicker adherends. Welding with a higher force, which is known to reduce hammering, corroborated this hypothesis as fiber squeeze-out within the top adherend was not observed. It is believed that hammering contributes to heating by causing an oscillatory impact excitation that is close to the natural frequencies of the system, which would result in amplification of the cyclic strain and subsequent increase in the viscoelastic heating in the adherend.

Keywords: fusion bonding; ultrasonic welding; thermoplastic composites; adherend thickness; hammering effect

Citation: Guevara-Sotelo, N.S.; Fernandez Villegas, I. Effect of Adherend Thickness on Near-Field Ultrasonic Welding of Single-Lap CF/LMPAEK Thermoplastic Composite Joints. *Materials* **2023**, *16*, 6968. https://doi.org/10.3390/ma16216968

Academic Editors: Patricia Krawczak, Chung-Hae Park and André Chateau Akué Asséko

Received: 20 September 2023
Revised: 25 October 2023
Accepted: 27 October 2023
Published: 30 October 2023

Copyright: © 2023 by the authors. Licensee MDPI, Basel, Switzerland. This article is an open access article distributed under the terms and conditions of the Creative Commons Attribution (CC BY) license (https://creativecommons.org/licenses/by/4.0/).

1. Introduction

Ultrasonic welding is a very interesting welding technique for thermoplastic composite assemblies mainly due to its ultra-fast heating rates and its ease of automation. Ultrasonic welding of thermoplastic composites is based on the application of high-frequency, typically 20 kHz, and low-amplitude mechanical vibrations perpendicular to the welding interface [1–3]. A sonotrode connected to a piezoelectric converter through a booster and to a press is used to transmit the vibrations into the material as well as to apply a certain static welding pressure throughout the welding process. Heat is generated through a combination of surface and viscoelastic friction. An energy director, in the form of resin-rich protrusions, a simple resin film, or a discontinuous resin film [4], is placed at the welding interface to ensure preferential heat generation at that location. As the vibrations are introduced in the welding stack (i.e., adherends and energy director), the energy director heats up, melts, and is (partially) squeezed out. Figure 1 shows such a succession of events for a discontinuous energy director. As a result, wetting of the adherends by the energy director occurs, which is a necessary condition for molecular inter-diffusion to ensue. As the final step, also known as the consolidation stage, the vibrations are stopped, and the weld is allowed to cool down under pressure. Ultrasonic welding is by nature a spot welding technique which, when applied sequentially, allows to create multi-spot welded overlaps [5,6]. When introducing a continuous relative movement between the welding and the parts to be joined (i.e. adherends), continuous ultrasonically welded overlaps can also be obtained [4,7,8].

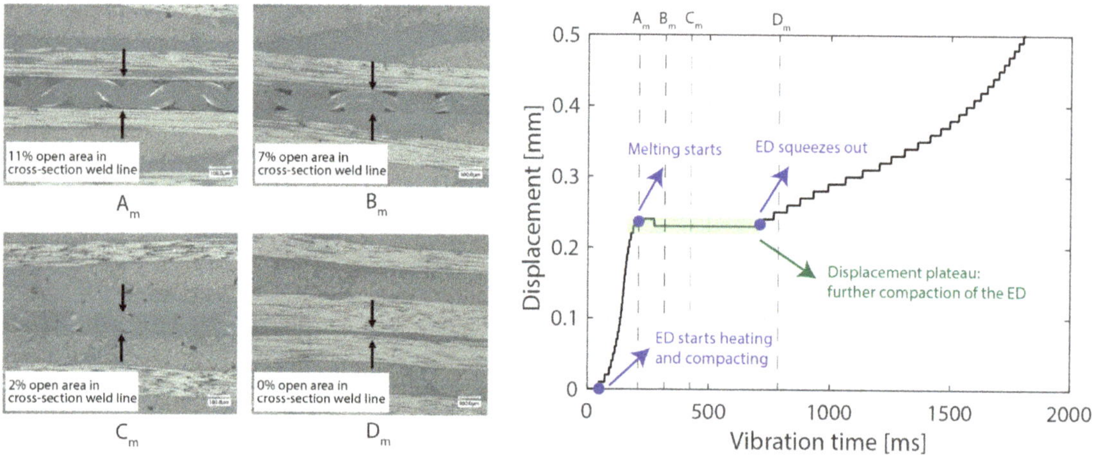

Figure 1. Cross-section micrographs showing the different steps a discontinuous energy director goes through under the ultrasonic vibrations, i.e., heating and further compaction followed by melting and squeeze-out (**left**). The same events can be identified in the downward displacement of the sonotrode (**right**). Adapted from [4].

In the last decade, there has been a renewed interest in the ultrasonic welding process, as indicated by the publication of numerous scientific articles on different aspects such as the impact of the morphology of the energy director [9,10], of the process parameters [11,12], and even of misaligned adherends [13,14] in the welding process. Interestingly, most of the research results reported in the open literature about ultrasonic welding of thermoplastic composites do not consider the thickness of the adherends as a variable. However, knowledge of the effect of the adherend thickness on the process and, especially, on its limits is of utmost importance for the selection of future applications for this welding technology. In the current state of the art, the thickness of the adherends is typically around 2 mm. In a previous paper by Fernandez Villegas on ultrasonic welding of carbon-fiber-reinforced polyetherimide composites [12], the author, however, doubled the thickness of the adherends (from 1.92 to 3.84 mm) as a side study to check the validity of some of the results obtained with 1.96 mm thick adherends concerning the use of the downward displacement of the sonotrode to control the weld quality. The results showed that the displacement value resulting in high-quality welds was the same in both cases; however, the overall power and energy consumed in the process were higher in the case of the thicker adherends.

Contrarily, the topic of adherend thickness, or, more precisely, the distance between the tip of the sonotrode and the welding interface (L), has been discussed in greater measure in relation to ultrasonic welding of unreinforced thermoplastics. Based on that, two "types" of welding processes, near-field and far-field welding, have been defined based on whether L is lower or higher than a certain threshold value, respectively. Said threshold is considered to be 6 mm for usual thermoplastics, with wavelengths (λ) between 60 and 130 mm at 20 kHz [15,16]. During near-field welding, the amplitude of the vibrations reaching the welding interface can be considered similar to the amplitude of the vibrations exerted by the sonotrode. Contrarily, in far-field welding, the amplitude of the vibrations at the welding interface depends on wave propagation through the material, which is affected by the ratio between L and λ as well as by wave attenuation [1,17]. Consequently, far-field welding introduces additional requirements for the welding process to be successful (i.e., to produce similar results to near-field welding). This is especially the case when it comes to far-field welding of semi-crystalline plastics, which have higher energy requirements than amorphous plastics to be welded [1,17]. In particular, successful far-field welding of semi-crystalline thermoplastics requires the top adherend to be designed so that L is an integer multiple of 0.5λ. This ensures a displacement antinode and, hence, maximum heat generation at the energy director rather than at the interface between the sonotrode and the top adherend [1,17]. Given the typical λ values mentioned above (between 60 and 130 mm), this size requirement can be practical in, for instance, butt-welding of cylindrical hollow parts [17] but not so practical in, for instance, overlap-welding of thermoplastic composite laminates (assuming similar wavelengths in reinforced thermoplastics).

This paper aims to provide initial insight into the effect of the adherend thickness on near-field ultrasonic welding of thermoplastic composites in a single-lap configuration. The material of choice for this research is carbon-fiber-reinforced low-melting polyaryletherketone (CF/LMPAEK), a semi-crystalline composite material of interest for high-performance, e.g., aerospace, applications. The research focuses on near-field ultrasonic welding given the typically thin-walled nature of high-performance thermoplastic composite structures. Based on previous experiments [12], it is expected that increasing the thickness of the adherends will, at the very least, increase the power consumed during the process, which might limit their applicability before the far-field threshold is reached. To look into this matter, the impacts of varying the thickness of the top adhered and of the bottom adherend were separately investigated. The regular output of the ultrasonic welding machine in terms of consumed power, downward displacement of the sonotrode, and amplitude was used as a first approach to observe changes in the welding process. Temperature measurements at the welding interface and within the adherends as well as high-speed camera recordings during the welding process were used to gain more insight into said changes.

2. Materials and Methods

2.1. Materials

The composite material used for the adherends was TC1225 CF/LMPAEK provided by Toray Advanced Composites (The Netherlands). The material featured a five-harness satin T300JB woven carbon fiber reinforcement with 277 g/m^2 fiber areal weight, 42% resin volume content, and nominal 0.31 mm consolidated ply thickness. The LMPAEK resin had a glass transition temperature of 147 °C and a melting temperature of 305 °C. Further, 580 mm × 580 mm composite laminates were manufactured by consolidating stacks of powder-impregnated composite layers in a hot platen Joos press (Pfalzgrafenweiler, Germany) at 365 °C and 10 bar for 30 min. Heating and cooling rates were set at 7 °C/min. For the thicker adherends, temperatures were checked during the consolidation cycle by placing K-type thermocouples (GG220-2K-0 provided by Tempco B.V., Bodegraven, The Netherlands) at the edges of the laminate (middle plane). The quality of the composite laminates was evaluated by C-scan (Olympus EPOCG 650, Hoofddorp, The Netherlands). Finally, the laminates were cut into single-lap shear adherends measuring 25.4 mm × 101.6 mm using a water-cooled Proth grinding machine (Taiwan). A 0.5 mm thick discontinuous LMPAEK film provided by Victrex (Middlesbrough, UK) was used for the energy directors. These discontinuous films present open areas. The energy directors were cut into rectangles slightly larger than the welding overlap (25.4 mm × 12.7 mm). Before welding, both the adherends and the energy directors were cleaned with a degreasing agent Hysol QD (PT Technologies Europe, Watergrasshill, Ireland).

Assuming the threshold between near-field and far-field welding to be at $L = 6$ mm (with L being the distance between the tip of the sonotrode and the welding interface, i.e., approximately the thickness of the top adherend), the adherend thicknesses considered in this study ranged from 1.17 mm (for a 4-ply laminate) to 5.79 mm (for a 20-ply laminate). Table 1 shows all the different thicknesses as well as laminate architectures used. The above-mentioned assumption was based on the fact that the wavelength of sound for the

CF/LMPAEK material used in this study was measured to be 146 mm at 20 kHz (time of flight measurements on pulse-echo ultrasonic inspection, Olympus Omniscan MX with phased array technology). This is not very different from typical values provided in the literature for which the threshold between near field and far field is considered to be 6 mm [15,16].

Table 1. Thicknesses and laminate architectures used in this work.

Nominal Thickness (mm)	Laminate Architecture	Number of Plies
1.17	$[(0/90)_2]_s$	4
1.83	$[(0/90)_3]_s$	6
2.37	$[(0/90)_4]_s$	8
3.55	$[(0/90)_6]_s$	12
4.72	$[(0/90)_8]_s$	16
5.79	$[(0/90)_{10}]_s$	20

2.2. *Welding*

A VE20 Slimline dialog 6200 ultrasonic welder from Herrmann Ultrasonics (Karlsbad, Germany) with a frequency of 20 kHz and 6.2 kW maximum power was used in this study. The welder was equipped with a rectangular sonotrode with a contact area of 15 mm × 30 mm and a gain of 1:1.7, and a booster with a gain of 1:2. With this setup, the welder can deliver a maximum peak-to-peak amplitude of 86.2 μm and a force between 130 and 2500 N. As the generic output of the ultrasonic welding process, the welder provides information about the power consumed during the process, and the actual amplitude and vertical displacement of the sonotrode.

For the experiments performed in this study, the welding amplitude was set to 80 μm and the welding force to 500 N. These were chosen as a starting point based on typical force and amplitude values used in previous studies on ultrasonic welding of thermoplastic composites. Such values are not based on previous optimizations as the goal of this study was to gain an understanding of the effect of changing the thickness, and not to optimize the joint strength or other parameters. The duration of the vibration was either directly controlled (time control) or indirectly controlled through the downward displacement of the sonotrode. For the latter, the target displacement was set to 0.5 mm, i.e., the thickness of the energy director. This is further referred to as a "full weld". It is known that such a long target displacement is beyond the point where high-quality welds are obtained [4,8], but, since this research did not aim at weld optimization, the long target displacement was chosen to provide a wide view of the welding process. The sonotrode was removed immediately after the vibration, meaning that there was no consolidation phase present in this study to avoid further material squeeze-out due to the applied pressure.

The clamping jig used is shown in Figure 2. It consists of a base, bar clamps, and pins for positioning the adherends and it is designed to weld single-lap shear specimens in which the adherends have the dimensions 25.4 mm × 101.6 mm and the overlap is 25.4 mm × 12.7 mm. To ensure parallelism between the bottom and top adherends, a dummy adherend (with a thickness equal to that of the bottom adherend) and an energy director were placed below the top adherend. A summary of the experiments conducted in this study is presented in Table 2 (changing the top adherend's thickness) and Table 3 (changing the bottom adherend's thickness). It is important to note that the thickness of the bottom adherend was kept constant (1.83 mm) when changing the top adherend's thickness and vice versa. The control parameter time was used only to stop the weld at a certain position in the displacement curve and it is not based on any previous optimization. Note that, for the case of repetitions with thermocouples, in some cases, the thermocouples failed, and the readings are therefore not presented in the results.

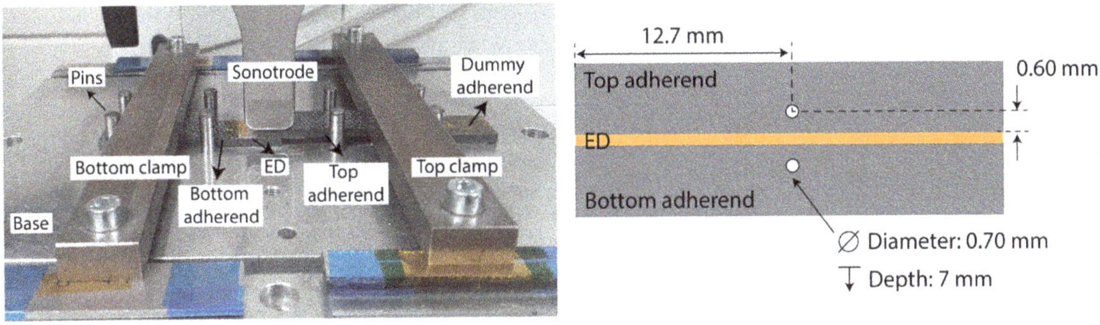

Figure 2. (**Left**): welding setup. (**Right**): schematic indicating thermocouple positioning.

Table 2. Summary of the experiments presented in this study for the change in the top adherend's thickness. The bottom adherend's thickness was kept constant (1.83 mm). The experiments with thermocouples had all thermocouples present (interface, top adherend, and bottom adherend). T: top adherend's thickness, TCs: thermocouples, d: displacement, t: time. Five repetitions (rep.) with TCs were completed for each case; however, only the repetitions in which the TCs survived are reported in the table.

T [mm]	Force [N]	Amplitude [μm]	Rep. without TCs	Rep. with TCs	Control	Purpose of the Experiments
1.17	500	80	3	1	d: 0.50 mm	To obtain power, displacement, amplitude, and temperature behavior during a full weld for comparison between the different thicknesses
1.83	500	80	3	1		
2.37	500	80	3	1		
3.55	500	80	3	2		
4.72	500	80	3	1		
3.55	500	80	3	0	t: 250 ms	To obtain micrographs at different positions of the displacement curve
3.55	500	80	3	0	t: 600 ms	
3.55	500	80	3	0	t: 800 ms	
1.17	1500	80	3	0	d: 0.50 mm	To study the effect of the thickness with a higher force value
3.55	1500	80	3	0		
4.72	1500	80	3	0		

Table 3. Summary of the experiments presented in this study for the change in the bottom adherend's thickness. The top adherend's thickness was kept constant (1.83 mm). The welding parameters were 500 N and 80 μm. The experiments with thermocouples had all thermocouples present (interface, top adherend, and bottom adherend). B: bottom adherend's thickness, TCs: thermocouples, d: displacement, t: time. Five repetitions (rep.) with TCs were completed for each case; however, only the repetitions in which the TCs survived are reported in the table.

B [mm]	Rep. without TCs	Rep. with TCs	Control	Purpose of the Experiments
1.17	3	2	d: 0.50 mm	To obtain power, displacement, amplitude, and temperature behavior during a full weld for comparison between the different thicknesses
1.83	3	1		
2.37	3	3		
3.55	3	0		
4.72	3	2		
5.79	3	1		

2.3. Analysis

The power consumed, the downward displacement of the sonotrode, and the actual amplitude of vibration during the vibration phase of the welding process were directly obtained from the ultrasonic welder.

Temperatures were measured at the welding interface and within the adherends using K-type thermocouples (GG220-2K-0, Tempco B.V., Bodegraven, The Netherlands) with a wire diameter of 0.10 mm and encapsulated diameter of 0.70 mm. For the temperature measurements within the adherends, holes were drilled in the adherends according to the schematic in Figure 2. The distance between the thermocouple hole and the interface (0.60 mm) was limited by the thickness of the thinner adherend (1.17 mm), and it was kept constant for all adherends to measure the temperature at the same position from the interface. For the temperature measurements at the welding interface, the measuring tip of the thermocouple was placed in the center of the overlap, on top of the bottom adherend and under the energy director. The thermocouples were connected to an analogue amplifier and sampled at 1 kHz. The temperature measurements were performed with three thermocouples (bottom adherend, interface, and top adherend). All thermocouples were placed simultaneously, which can affect the response of the welding process. However, because this added effect of placing thermocouples is present in all experiments when comparing different temperature evolutions for all thicknesses, it was not considered worrisome for this research. The raw thermocouple data were filtered in Matlab using a moving average filter with a 20-point window.

A FASTCAM NOVA S Series (Photron, Reutlingen, Germany) high-speed camera was used to record the welding process in slow motion using a rate of 1000 frames per second and a shutter speed of 1 μs. The camera was mounted on a tripod and positioned in such a way that the focus was on the middle of the welding stack. One extra lamp was required to achieve the desired lighting for the provided recording parameters.

As was mentioned earlier in the Introduction, the effects of changing the thickness of the top adhered and the bottom adherend were studied separately. Therefore, the welded joints will have a significant thickness mismatch between the top and bottom adherends. Because of this, it was decided not to perform any lap shear testing. Testing joints with different thicknesses and plainly comparing these values could be misleading because new variables, such as the bending stiffness which varies with thickness, are now involved.

3. Results

Only results are presented in this section. The analysis of the data, the trends, and further explanations are presented in the Discussion.

3.1. Thickness of the Top Adherend

Figure 3 shows the power consumed by the ultrasonic welder and the vertical downward displacement undergone by the sonotrode for five different thicknesses of the top adherend (ranging from 1.17 mm to 4.72 mm) and a fixed thickness of the bottom adherend (1.83 mm). One representative example is shown per thickness value. It should be noted that, due to a persistent error in the ultrasonic welder (maximum amplitude value exceeded), it was not possible to weld the configuration with a 5.79 mm thick top adherend, as originally intended. Figure 4 illustrates the repeatability of the results for two of those five thicknesses (1.17 mm and 3.55 mm). Figure 5 shows the amplitude as measured by the ultrasonic welder for three thicknesses of the top adherend with the corresponding sonotrode displacement curve superimposed for reference. Figure 6 shows temperature measurements at the welding interface within the bottom adherend and the top adherend for five different thicknesses and two repetitions for a top adherend thickness of 3.55 mm. Figure 7 gathers a selection of snapshots from the high-speed camera at three different stages in the welding process for two thicknesses of the top adherend, 2.37 mm and 3.55 mm.

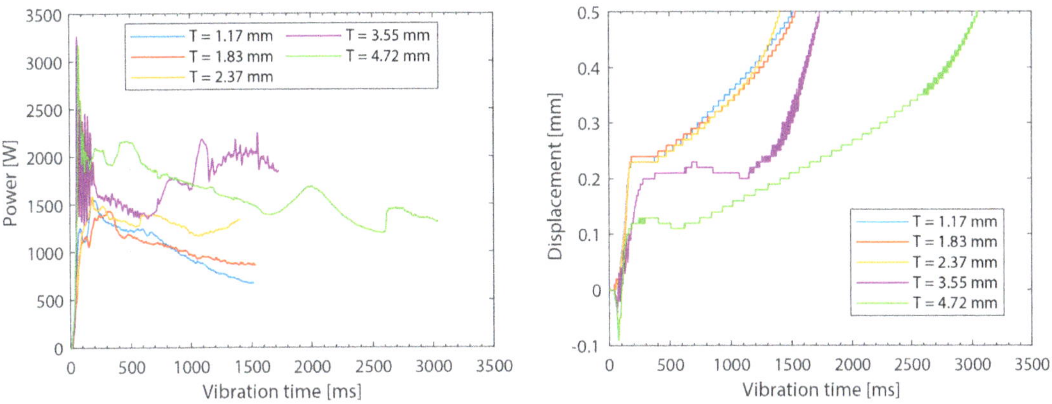

Figure 3. Consumed power (**left**) and downward displacement (**right**) curves for top adherend with varying thickness and discontinuous energy directors. One representative curve per thickness value. The welding parameters were 500 N force and 80 µm vibration amplitude. The bottom adherend's thickness was 1.83 mm.

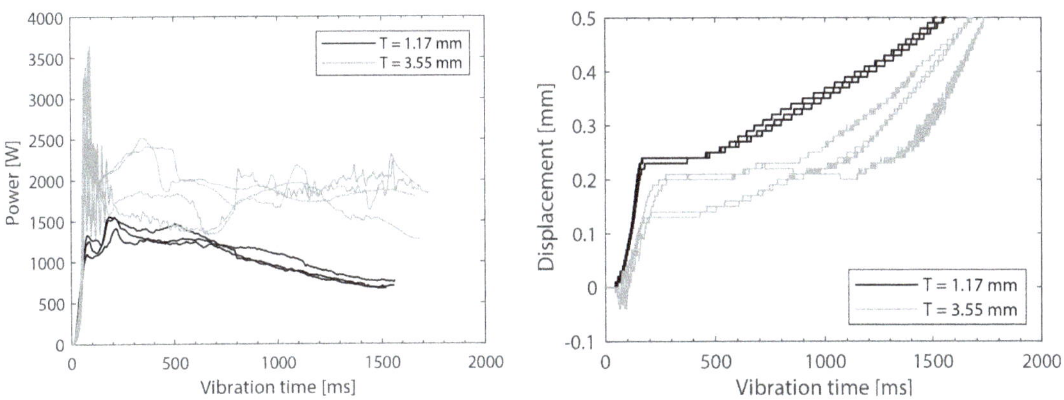

Figure 4. Consumed power (**left**) and downward displacement (**right**), three repetitions for two top adherend thickness values (1.17 mm and 3.55 mm). The welding parameters were 500 N force and 80 µm vibration amplitude. The bottom adherend's thickness was 1.83 mm.

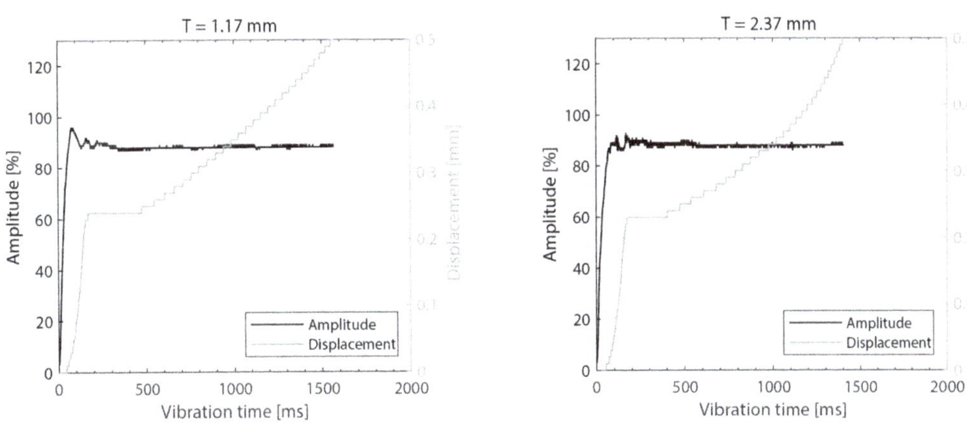

Figure 5. *Cont.*

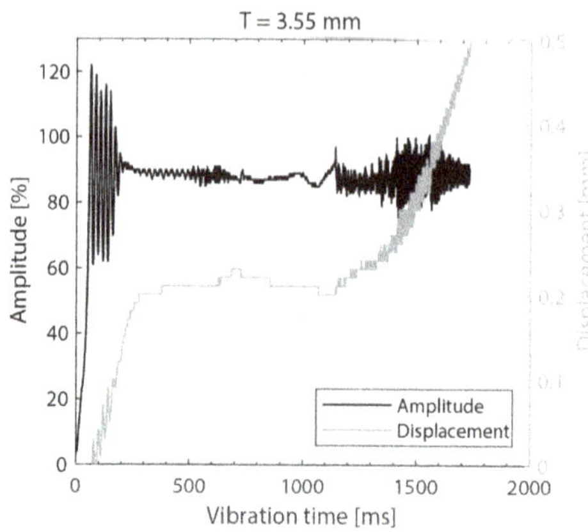

Figure 5. Amplitude and downward displacement curves for different thicknesses of the top adherend: 1.17 mm (**top-left**), 2.37 mm (**top-right**), and 3.55 mm (**bottom**). The welding parameters were 500 N force and 80 µm vibration amplitude. The bottom adherend's thickness was 1.83 mm. The amplitude is provided as a percentage of the maximum amplitude the equipment can deliver.

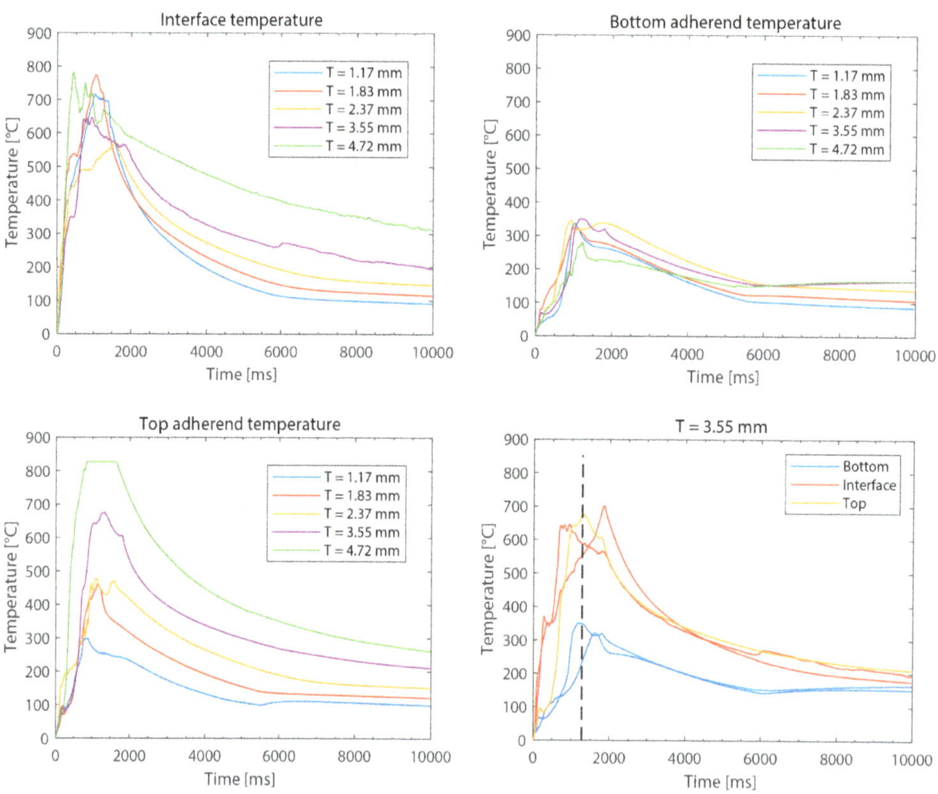

Figure 6. Representative temperature curves for different thicknesses of the top adherend: interface temperature (**top left**), bottom adherend temperature (**top right**), top adherend temperature (**bottom left**). (**Bottom right**): two repetitions for a top adherend thickness of 3.55 mm. For the temperature in the top adherend, only one thermocouple survived. The dashed line represents an estimate of the end of the vibrations. The welding parameters were 500 N force and 80 µm vibration amplitude. The bottom adherend's thickness was 1.83 mm.

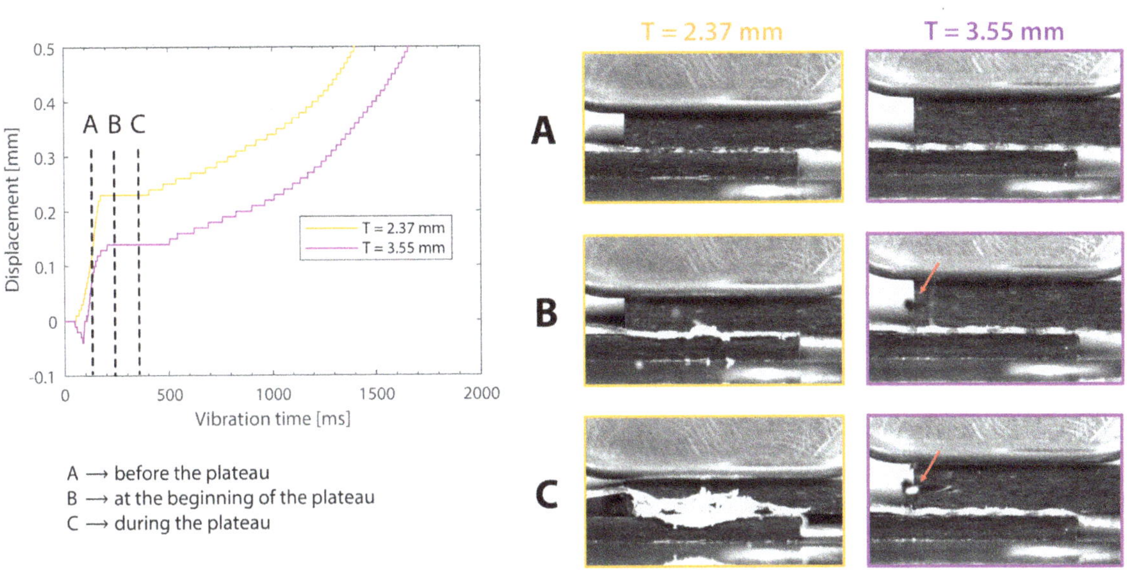

Figure 7. High-speed camera snapshots at different times in the welding process ((**A**) before the displacement plateau, (**B**) at the beginning of the displacement plateau, (**C**) during the displacement plateau) for two thicknesses of the top adherend (2.37 mm and 3.55 mm). The red arrows indicate fiber squeeze-out. The welding parameters were 500 N force and 80 μm vibration amplitude. The bottom adherend's thickness was 1.83 mm.

3.2. Thickness of the Bottom Adherend

Figures 8–12 show the same data as shown above but for the experiments in which the thickness of the bottom adherend was varied. In this case, it was possible to obtain results from the six thickness values originally planned, ranging from 1.17 mm to 5.79 mm.

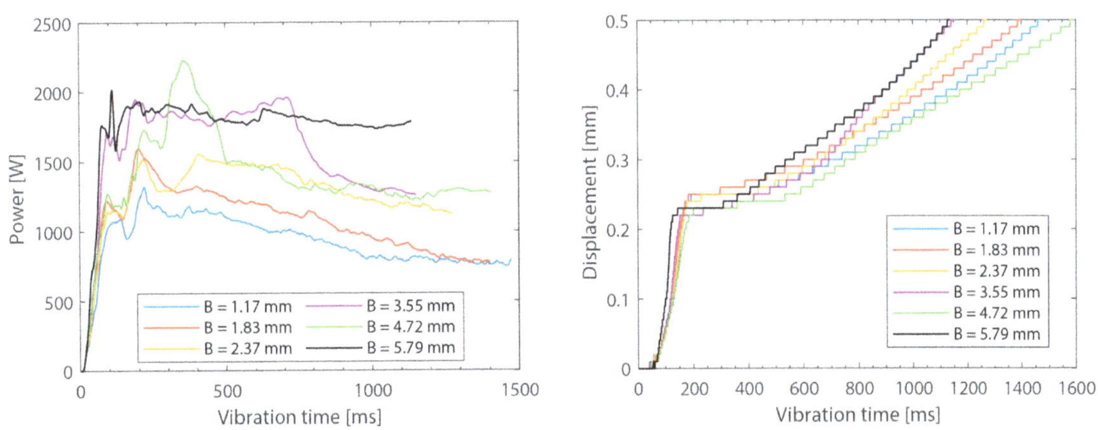

Figure 8. Consumed power (**left**) and downward displacement (**right**) curves for bottom adherend with varying thickness and discontinuous energy directors. One representative curve per thickness value. The welding parameters were 500 N force and 80 μm vibration amplitude. The top adherend's thickness was 1.83 mm.

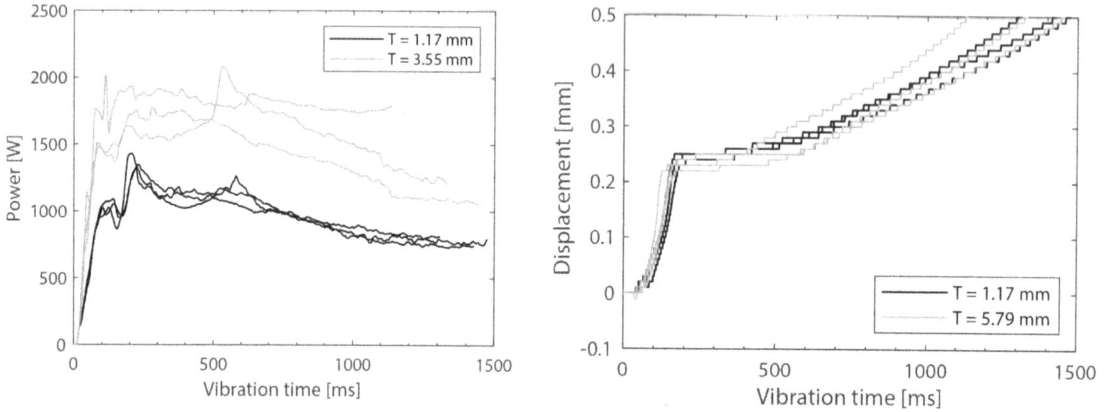

Figure 9. Consumed power (**left**) and downward displacement (**right**), three repetitions for two bottom adherend thickness values. The welding parameters were 500 N force and 80 μm vibration amplitude. The top adherend's thickness was 1.83 mm.

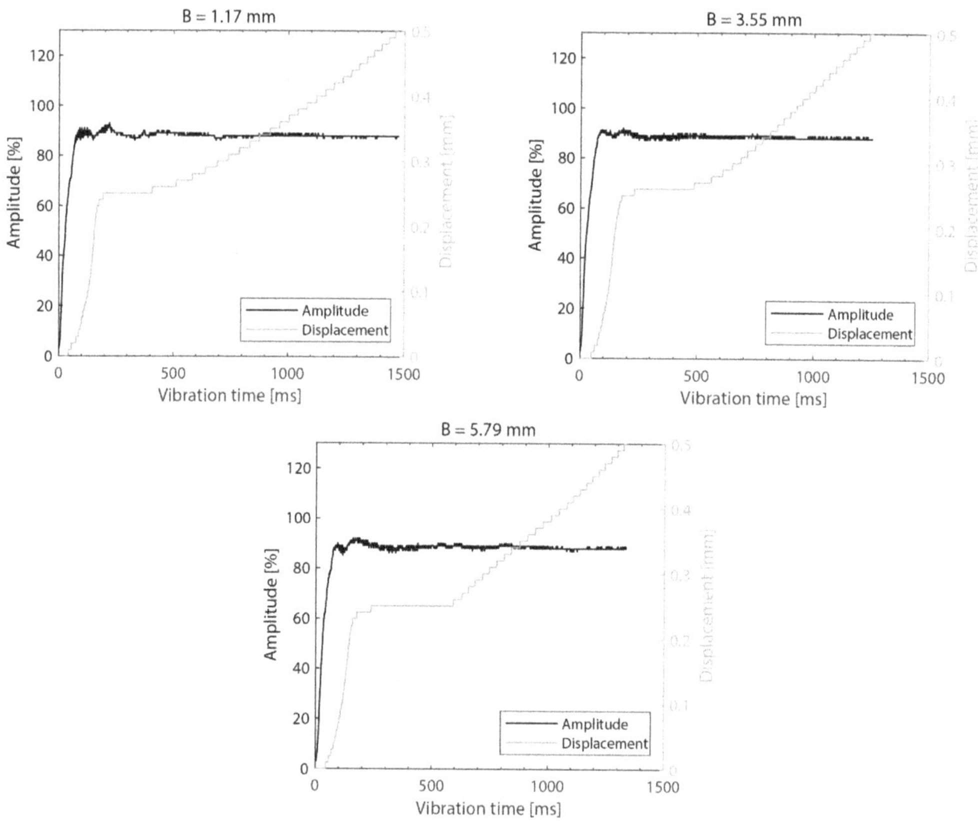

Figure 10. Amplitude and displacement curves for different thicknesses of the bottom adherend: 1.17 mm (**top-left**), 3.55 mm (**top-right**), and 5.79 mm (**bottom**). The welding parameters were 500 N force and 80 μm vibration amplitude. The top adherend's thickness was 1.83 mm. The amplitude is provided as a percentage of the maximum amplitude the equipment can deliver.

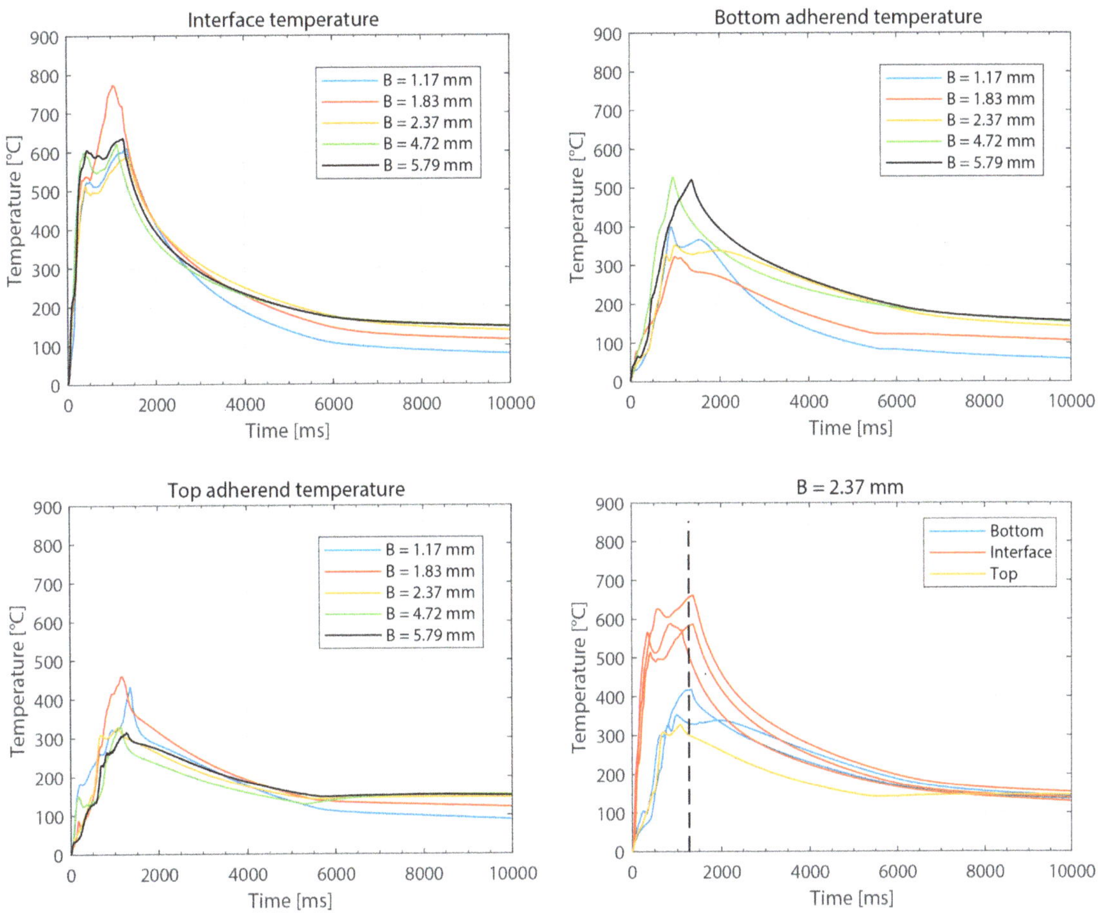

Figure 11. Representative temperature curves for different thicknesses of the bottom adherend: interface temperature (**top left**), bottom adherend temperature (**top right**), top adherend temperature (**bottom left**). Note that the temperature readings for 3.55 mm—thick bottom adherend are not present in these graphs since the decision to include this thickness in the study was made after the temperature measurements were performed. (**Bottom right**): three repetitions for a bottom adherend thickness of 2.37 mm. For the temperature in the top adherend, only one thermocouple survived, and, for the temperature in the bottom adherend, only two survived. The dashed line represents an estimate of the end of the vibrations. The welding parameters were 500 N force and 80 µm vibration amplitude. The top adherend's thickness was 1.83 mm.

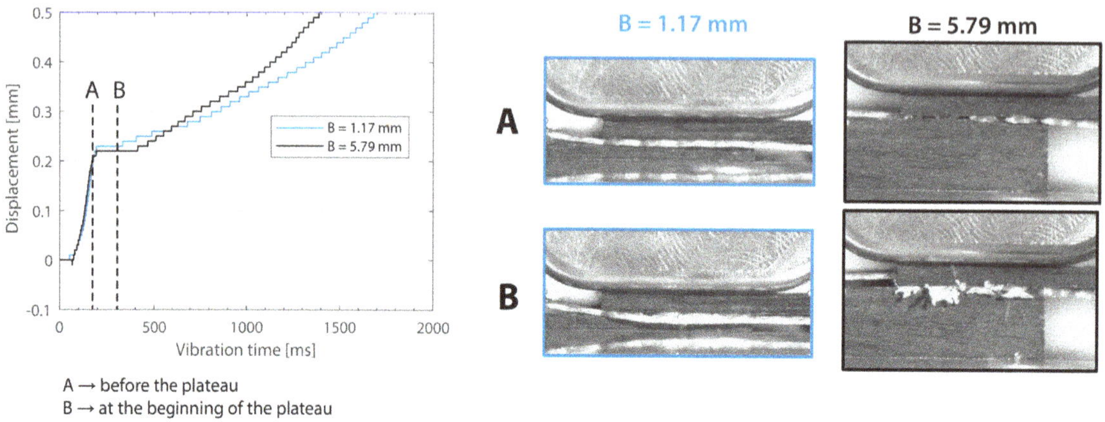

Figure 12. High-speed camera snapshots at different times in the welding process ((**A**) before the displacement plateau, (**B**) during the displacement plateau) for two thicknesses of the bottom adherend (1.17 mm and 5.79 mm). The welding parameters were 500 N force and 80 μm vibration amplitude. The top adherend's thickness was 1.83 mm.

4. Discussion

This study aimed to gain an initial understanding of the effect of changing the thickness of the adherends during near-field ultrasonic welding of thermoplastic composites. The behavior of the process and the evolution of the welded joints when varying the thicknesses of the top and bottom adherends were studied using output data from the ultrasonic welder, temperature measurements, and high-speed camera recordings.

As expected, increasing the thickness of the adherends increased the overall power consumption during the welding process. This effect was observed both when increasing the thickness of the top adherend (Figure 3) as well as when increasing the thickness of the bottom adherend (Figure 8). The main difference between the two cases was that increasing the thickness of the top adherend did cause a significant overshoot of the power peak observed at the onset of the process (Figure 3). This phenomenon may have been the reason why it was not possible to operate the ultrasonic welder for the highest top adherend thickness considered in this work, i.e., 5.79 mm. The observed differences may be related to the differences in wave transmission between the top and the bottom adherend. Indeed, every interface between two different media, i.e., sonotrode–top adherend, top adherend–energy director, and energy director–bottom adherend, will reflect a portion of the waves, resulting in a decrease in the vibrations transmitted across each interface. Consequently, it is sensible to assume that the process will be more sensitive to thickness changes in the top adherend since it receives the largest share of the vibration energy. It is, however, interesting to note that the thickness of the bottom adherend affects the power consumed throughout the process, even though the transmission of vibrations across the welding interface is said to significantly decrease once the energy director is molten [18].

Beyond this, increasing the thickness of the top adherend was found to affect the downward displacement of the sonotrode during the welding process. As shown in Figure 1, the vertical displacement of the sonotrode is usually linked to the physical changes occurring at the welding interface during the welding process. In the case of a discontinuous energy director, as is the case in this study, the sonotrode will first travel downwards as it compresses the energy director until its original empty spaces are filled out with material [4]. At that point, the sonotrode will remain stationary (displacement plateau) until the energy director is molten and starts to flow, which will cause further downward displacement of the sonotrode (see Figure 1). All the curves in Figure 4 show roughly this behavior; however, it is interesting that, whereas for the 2.37 mm thick adherend and below, all the displacement curves overlap, for 3.55 mm and 4.72 mm thick adherends, the displacement curves show greater variability. One remarkable feature of the displacement curves corresponding to the thicker adherends is that the displacement plateau is reached at a lower downward displacement, which is surprising considering that both the energy director and the welding parameters are the same in all cases. The high-speed camera images in Figure 7 help to explain this behavior since they show that, in the case of the thicker top adherend, damage in the form of melting and fiber squeeze-out in the top adherend take place during the displacement plateau, as opposed to just melting of the energy director, as is the case for the thinner adherend. Temperature measurements (Figure 6) confirm that observation since they show a significant increase in the temperatures within the top adherend as its thickness increases. It should be noted that, due to potential errors in the readings caused by the interaction between the thermocouples and the high-frequency vibrations [19], the temperature data were only considered from a qualitative viewpoint. It is interesting to point out that, even though the images in Figure 7 do not clearly indicate melting of the energy director and hence the creation of a welded joint within the displacement plateau for the thicker adherend, the temperature readings at the welding interface seem to indicate the contrary since they are similar for the thin and the thick adherends (Figure 6). Cross-section micrographs of welds allowed to cool down without any consolidation pressure settle this argument by showing melting of the energy director (in the form of deconsolidation voids) within the plateau also in the cases of the thicker adherends (Figure 13). None of the effects described in this paragraph regarding heating of the adherends were, however, observed when increasing the thickness of the bottom adherend, as shown in Figures 9, 11 and 12.

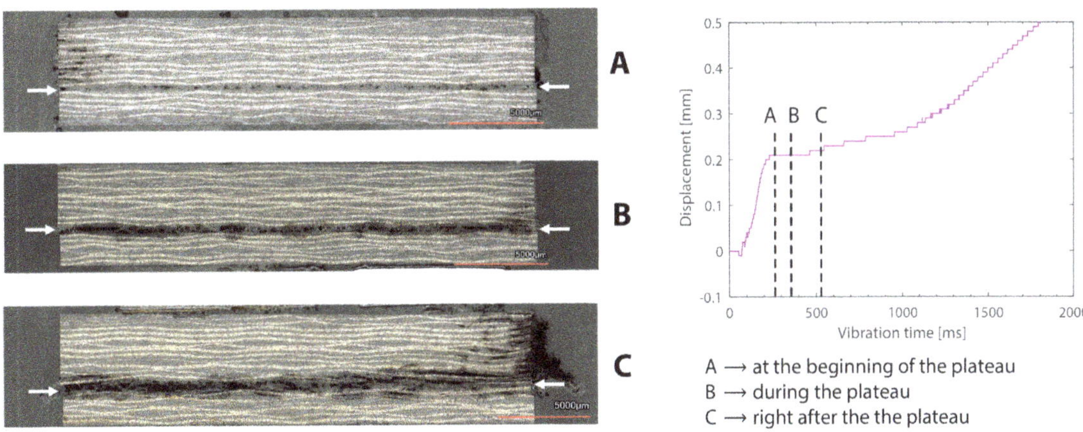

Figure 13. Cross-section micrographs of welded joints allowed to cool down without any consolidation pressure at different times in the welding process ((**A**) at the beginning of the displacement plateau, (**B**) during the displacement plateau, (**C**) right after the displacement plateau) for a top adherend thickness of 3.55 mm. The white arrows indicate the weldline. The welding parameters were 500 N force and 80 μm vibration amplitude. The bottom adherend's thickness was 1.83 mm. Note that there is porosity observed at the weldline, which can be attributed to the absence of a consolidation phase. Note that the criteria for the micrographs were according to the position in the displacement curve, which is the reason why the times indicated in the graph do not match with the times provided in Table 2.

The amplitude readings in Figures 5 and 10 suggest the existence of significant amplitude oscillations, which were identified as hammering, in the case of the 3.55 mm thick top adherend. It should be noted that, of all the different thicknesses shown in those figures, that is the only case in which the displacement curve presents an "anomalous" behavior (Figure 3). Hammering is an unavoidable phenomenon in ultrasonic welding caused by the periodic loss of contact between the sonotrode and the top adherend [20]. To obtain an indication of whether hammering could be contributing to the temperature increase and melting observed within the thicker top adherends, an extra set of experiments with 1.17 mm, 3.55 mm, and 4.72 mm thick top adherends and a tripled welding force (1500 N) were carried out. It should be noted that increasing the welding force is known to improve the contact between the sonotrode and the top adhered [17], thereby reducing hammering [21], as also confirmed by the amplitude graphs in Figure 14 (see Figure 5 for reference). The new displacement curves (Figure 15) show, unlike those in Figure 4, overlapped displacement plateaus. Consistently, the high-speed camera images show no trace of melting and fiber squeeze-out within the adherends (Figure 16). Furthermore, a comparison of the surface of the top adherend, which is in contact with the sonotrode during the vibration phase, for both welding forces used in this study (Figure 17) shows the extent of the damage in the top adherend for the 500 N force case and the significant reduction in fiber squeeze-out and melting for the 1500 N case. Note that the observed fiber squeeze-out when welding with a 1500 N force is attributed to the fact that this corresponds to a full weld, which is known to be beyond the optimal point for joint strength in which fiber squeeze-out is expected to occur. These results are a clear indication of the role of hammering in the observed phenomena. Regarding the mechanism through which hammering might be contributing to the increased heating of the top adherend, the hypothesis is that it causes an oscillatory impact excitation, which, if close to the natural frequencies of the system, could result in an amplification of the cyclic strain and hence viscoelastic heating in the adherend. A precise modal analysis of the system as well as an analysis of the periodicity of hammering would be necessary to test such a hypothesis. Finally, the occurrence of more hammering in thicker adherends, which are nevertheless more compliant than thinner ones [19], can be explained by considering that, the more compliant a material is in static conditions, the more it behaves like a hard spring under dynamic loading [2].

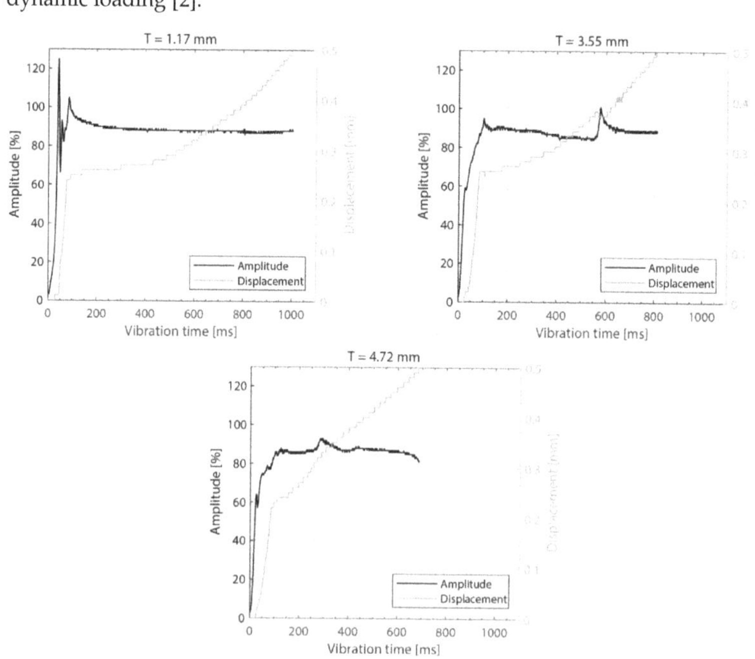

Figure 14. Amplitude and displacement curves for different thicknesses of the top adherend and increased welding force: 1.17 mm (**top-left**), 3.55 mm (**top-right**), and 4.72 mm (**bottom**). The welding parameters were 1500 N force and 80 µm vibration amplitude. The bottom adherend's thickness was 1.83 mm. The amplitude is provided as a percentage of the maximum amplitude the equipment can deliver.

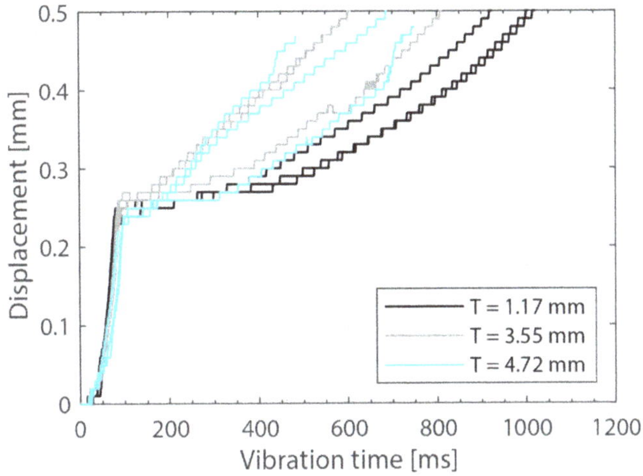

Figure 15. Displacement curves for increased welding force and different thicknesses of the top adherend (1.17 mm, 3.55 mm, and 4.72 mm). The welding parameters were 1500 N force and 80 µm vibration amplitude. The bottom adherend's thickness was 1.83 mm.

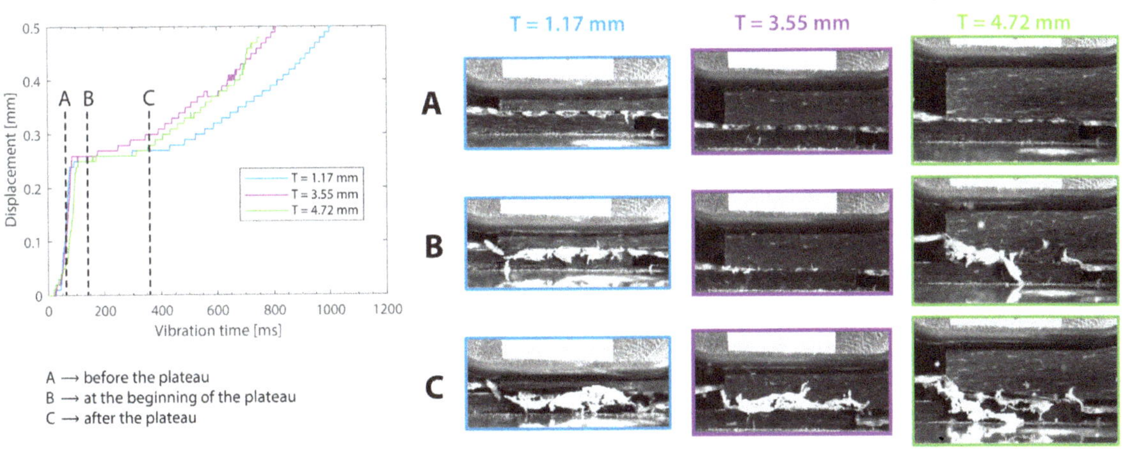

Figure 16. High-speed camera snapshots at different times in the welding process ((**A**) before the displacement plateau, (**B**) at the beginning of the displacement plateau, (**C**) after the displacement plateau) for three thicknesses of the top adherend (1.17 mm, 3.55 mm, and 4.72 mm). The welding parameters were 1500 N force and 80 µm vibration amplitude. The bottom adherend's thickness was 1.83 mm.

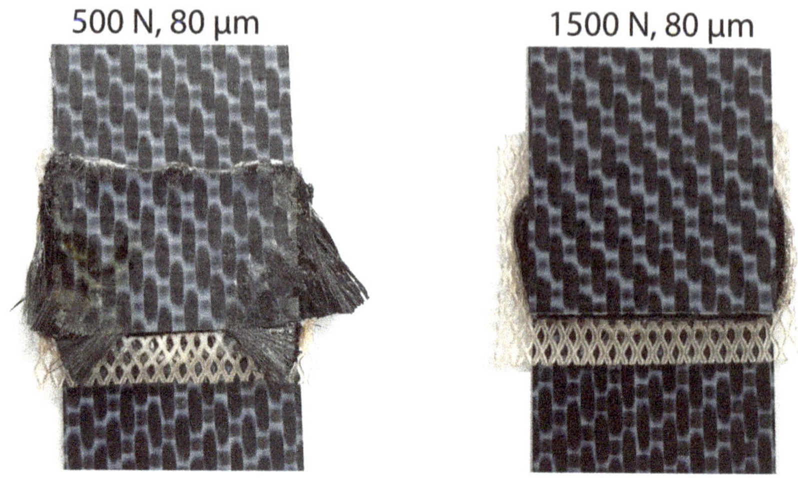

Figure 17. Comparison of the top adherend surface when welding with a force of 500 N (**left**) and 1500 N (**right**). This surface is in contact with the sonotrode during the vibration phase. The thickness of the top adherend was 3.55 mm, the thickness of the bottom adherend was 1.83 mm, and the welding amplitude was 80 μm. The joints were welded until a downward displacement of the sonotrode of 0.50 mm was reached.

5. Conclusions

The objective of this study was to provide an initial insight into the effect of the adherend thickness on the near-field static ultrasonic welding of fabric CF/LMPAEK thermoplastic composites. The results showed that increasing the thickness of the top adherend has a much more significant effect in the welding process than increasing the thickness of the bottom adherend. Indeed, increasing the thickness of the bottom adherend was only found to have an apparent effect on the overall consumed power. The increase in power caused by an increase in the thickness of the bottom adherend was, however, within the operating limits of the ultrasonic welder and did not cause any disruptions in the welding process (up to, at least, the maximum thickness considered in this study, i.e., 5.79 mm). Contrarily, increasing the top adherend thickness beyond 4.72 mm was found to preclude welding, likely due to the significant overshoot in the power required at the onset of the process. Furthermore, bulk heating in the top adherend during the welding process was found to increase as its thickness increased, likely as a result of increased hammering. Significantly increasing the welding force seemed to mitigate this issue while not having a negative effect on weldability. Further research on the implications of changing the thickness and process parameters on the joint strength is needed to gain more insight into some of the effects identified in this paper, e.g., increasing the welding force to mitigate negative effects when increasing the thickness of top adherend, and to determine processing windows for high-quality welds with thick adherends.

Author Contributions: Conceptualization, N.S.G.-S. and I.F.V.; methodology, N.S.G.-S. and I.F.V.; validation, N.S.G.-S. and I.F.V.; formal analysis, N.S.G.-S.; investigation, N.S.G.-S.; resources, I.F.V.; data curation, N.S.G.-S. and I.F.V.; writing—original draft preparation, N.S.G.-S. and I.F.V.; writing—review and editing, N.S.G.-S. and I.F.V.; visualization, N.S.G.-S. and I.F.V.; supervision, I.F.V.; project administration, N.S.G.-S.; funding acquisition, I.F.V. All authors have read and agreed to the published version of the manuscript.

Funding: This research received no external funding.

Institutional Review Board Statement: Not applicable.

Informed Consent Statement: Not applicable.

Data Availability Statement: Data repository available at https://doi.org/10.4121/0de8b885-73bd-480d-86c8-138af34c0c6c.v1 (accessed on 26 October 2023).

Acknowledgments: The authors would like to thank C.B.G. Brito for her contributions during the research and analysis.

Conflicts of Interest: The authors declare no conflict of interest.

References

1. Potente, H. Ultrasonic welding—Principles & theory. *Mater. Des.* **1984**, *5*, 228–234. [CrossRef]
2. Benatar, A.; Gutowski, T.G. Ultrasonic welding of PEEK graphite APC-2 composites. *Polym. Eng. Sci.* **1989**, *29*, 1705–1721. [CrossRef]
3. Yousefpour, A.; Hojjati, M.; Immarigeon, J.-P. Fusion Bonding/Welding of Thermoplastic Composites. *J. Thermoplast. Compos. Mater.* **2004**, *17*, 303–341. [CrossRef]
4. Jongbloed, B.; Teuwen, J.; Palardy, G.; Fernandez Villegas, I.; Benedictus, R. Continuous ultrasonic welding of thermoplastic composites: Enhancing the weld uniformity by changing the energy director. *J. Compos. Mater.* **2020**, *54*, 2023–2035. [CrossRef]
5. Zhao, T.; Broek, C.; Palardy, G.; Villegas, I.F.; Benedictus, R. Towards robust sequential ultrasonic spot welding of thermoplastic composites: Welding process control strategy for consistent weld quality. *Compos. Part A Appl. Sci. Manuf.* **2018**, *109*, 355–367. [CrossRef]
6. Zhao, T. Sequential Ultrasonic Spot Welding of Thermoplastic Composites. Ph.D. Thesis, Delft University of Technology, Delft, The Netherlands, 2018. [CrossRef]
7. Palardy, G.; Senders, F.J.M.; Villegas, I.F.; van Beurden, M. Continuous ultrasonic welding of thermoplastic composite plates. In Proceedings of the Automated Composites Manufacturing-Third International Symposium, Montreal, QC, Canada, 20–21 April 2017.
8. Jongbloed, B.; Teuwen, J.; Benedictus, R.; Villegas, I.F. On differences and similarities between static and continuous ultrasonic welding of thermoplastic composites. *Compos. Part B Eng.* **2020**, *203*, 108466. [CrossRef]
9. Fernandez Villegas, I.; Valle Grande, B.; Bersee, H.E.N.; Benedictus, R. A comparative evaluation between flat and traditional energy directors for ultrasonic welding of CF/PPS thermoplastic composites. *Compos. Interfaces* **2015**, *22*, 717–729. [CrossRef]
10. Palardy, G.; Villegas, I.F. On the effect of flat energy directors thickness on heat generation during ultrasonic welding of thermoplastic composites. *Compos. Interfaces* **2017**, *24*, 203–214. [CrossRef]
11. Tolunay, M.N.; Dawson, P.R.; Wang, K.K. Heating and bonding mechanisms in ultrasonic welding of thermoplastics. *Polym. Eng. Sci.* **1983**, *23*, 726–733. [CrossRef]
12. Villegas, I.F. Strength development versus process data in ultrasonic welding of thermoplastic composites with flat energy directors and its application to the definition of optimum processing parameters. *Compos. Part A Appl. Sci. Manuf.* **2014**, *65*, 27–37. [CrossRef]
13. Brito, C.B.G.; Teuwen, J.; Dransfeld, C.A.; Villegas, I.F. On improving process efficiency and weld quality in ultrasonic welding of misaligned thermoplastic composite adherends. *Compos. Struct.* **2023**, *304*, 116342. [CrossRef]
14. Brito, C.B.G.; Teuwen, J.; Dransfeld, C.A.; Villegas, I.F. The effects of misaligned adherends on static ultrasonic welding of thermoplastic composites. *Compos. Part A Appl. Sci. Manuf.* **2022**, *155*, 106810. [CrossRef]
15. Yeh, H.J. Ultrasonic welding of medical plastics. In *Joining and Assembly of Medical Materials and Devices*; Elsevier: Amsterdam, The Netherlands, 2013; pp. 296–323e. ISBN 978-1-84569-577-4.
16. Benatar, A. Ultrasonic welding of plastics and polymeric composites. In *Power Ultrasonics*; Elsevier: Amsterdam, The Netherlands, 2015; pp. 295–312. ISBN 978-1-78242-028-6.
17. Benatar, A.; Cheng, Z. Ultrasonic welding of thermoplastics in the far-field. *Polym. Eng. Sci.* **1989**, *29*, 1699–1704. [CrossRef]
18. Li, Y.; Liu, Z.; Shen, J.; Lee, T.H.; Banu, M.; Hu, S.J. Weld Quality Prediction in Ultrasonic Welding of Carbon Fiber Composite Based on an Ultrasonic Wave Transmission Model. *J. Manuf. Sci. Eng.* **2019**, *141*, 081010. [CrossRef]
19. Jongbloed, B.C.P.; Teuwen, J.J.E.; Benedictus, R.; Villegas, I.F. A Study on Through-the-Thickness Heating in Continuous Ultrasonic Welding of Thermoplastic Composites. *Materials* **2021**, *14*, 6620. [CrossRef] [PubMed]
20. Palardy, G.; Shi, H.; Levy, A.; Le Corre, S.; Fernandez Villegas, I. A study on amplitude transmission in ultrasonic welding of thermoplastic composites. *Compos. Part A Appl. Sci. Manuf.* **2018**, *113*, 339–349. [CrossRef]
21. Levy, A.; Le Corre, S.; Fernandez Villegas, I. Modeling of the heating phenomena in ultrasonic welding of thermoplastic composites with flat energy directors. *J. Mater. Process. Technol.* **2014**, *214*, 1361–1371. [CrossRef]

Disclaimer/Publisher's Note: The statements, opinions and data contained in all publications are solely those of the individual author(s) and contributor(s) and not of MDPI and/or the editor(s). MDPI and/or the editor(s) disclaim responsibility for any injury to people or property resulting from any ideas, methods, instructions or products referred to in the content.

materials

MDPI

Article

Experimental and Numerical Development on Multi-Material Joining Technology for Sandwich-Structured Composite Materials

Lucian Zweifel [1,*], Igor Zhilyaev [1], Christian Brauner [1], Martin Rheme [2], Gregor Eckhard [2], Valentin Bersier [2], Slobodan Glavaški [2] and Ricardo Pfeiffer [3]

[1] Institute of Polymer Engineering, FHNW University of Applied Sciences and Arts Northwestern Switzerland, Klosterzelgstrasse 2, 5210 Windisch, Switzerland; igor.zhilyaev@fhnw.ch (I.Z.); christian.brauner@fhnw.ch (C.B.)

[2] MultiMaterial-Welding AG, Zentralstrasse 115, 2503 Biel, Switzerland; martin.rheme@mm-welding.com (M.R.); gregor.eckhard@mm-welding.com (G.E.); valentin.bersier@mm-welding.com (V.B.); slobodan.glavaski@mm-welding.com (S.G.)

[3] KVT-Fastening AG (Bossard Group), Lagerstrasse 8, 8953 Dietikon, Switzerland; R.Pfeiffer@kvt-fastening.com

* Correspondence: lucian.zweifel@fhnw.ch

Citation: Zweifel, L.; Zhilyaev, I.; Brauner, C.; Rheme, M.; Eckhard, G.; Bersier, V.; Glavaški, S.; Pfeiffer, R. Experimental and Numerical Development on Multi-Material Joining Technology for Sandwich-Structured Composite Materials. *Materials* **2021**, *14*, 6005. https://doi.org/10.3390/ma14206005

Academic Editors: Patricia Krawczak, André Chateau Akué Asséko and Chung-Hae Park

Received: 30 August 2021
Accepted: 9 October 2021
Published: 12 October 2021

Publisher's Note: MDPI stays neutral with regard to jurisdictional claims in published maps and institutional affiliations.

Copyright: © 2021 by the authors. Licensee MDPI, Basel, Switzerland. This article is an open access article distributed under the terms and conditions of the Creative Commons Attribution (CC BY) license (https://creativecommons.org/licenses/by/4.0/).

Abstract: Creating connection points for sandwich-structured composites without losing technical performance is key to realising optimal lightweight structures. The patented LiteWWeight® technology presents cost-effective connections on sandwich panels in a fraction of a few seconds without predrilling. Ultrasonic equipment is used to insert a thermoplastic fastener into the substrate material and partially melt it into the porous internal structure. This creates a highly interlocked connection (connection strength is above 500 N) suitable for semi-structural applications. This study focused on the simulation and experimental validation of this process, mainly on the interaction between the pin and the substrate material during the joining process. The dynamic thermo-mechanical model showed reasonable agreement with experimental methods such as process data, high-speed camera monitoring or computed tomography and allowed the prediction of the connection quality by evaluation of the degree of interlock. The connection strength prediction by the developed model was validated within several various process setups, resulting in a prediction accuracy between 94–99% depending on the setup.

Keywords: joining; process modelling and simulation; sandwich structures; numerical analysis; ultrasonic technology; polymer composites

1. Introduction

The introduction of fibre-reinforced polymers in load-bearing automotive structures provides a great potential to significantly reduce weight, fuel consumption, and consequently CO_2 emissions. Reliable and cost-effective joining technologies must be developed to reduce production time and enable the manufacture and assembly of composite structural parts. Here, sandwich-structured composite materials provide clear advantages in terms of weight savings and mechanical performance in the aerospace and automotive industries owing to their high stiffness and strength to weight ratios. Such structures consist of two thin high density face sheets, bonded to a thick core made from low density foam material. Variation of the materials combination and foam to void ratio allows production of lightweight structural systems with desirable resulting material properties.

Using appropriate materials for appropriate scenarios is becoming more popular in transportation areas such as in the automotive and railway sectors, leading to multi-material design concepts with a high demand for reliable, cost-efficient joining methods [1,2]. Nevertheless, traditional fastening methods (metallic fasteners, adhesive bonding) for joining fibre-reinforced polymers decrease mechanical performance [3,4]. Metallic fasteners have a relatively high weight. Additionally, their thermal expansion coefficients and thermal conductivities are higher than those of the surrounding composite structure which could lead to composite degradation in the joint region with changes in the surrounding temperature. Furthermore, some metallic fasteners require holes to be drilled in advance, which could increase the chance of crack initiation in the joint area. Damage in the connection area produced by the conventional joining methods and further intensive loading conditions might debone the skin from the core and initiate the crack propagation [5,6].

Adhesive bonding is an alternative to metallic fasteners. However, these techniques strongly depend on the contacting surfaces morphology and presence of contaminants Bonding processes are time-consuming and difficult to analyse once cured. Moreover, debonding leads to significant damage to the contact region [7,8]. Fibre-reinforced thermoplastic rivets have recently been implemented for joining composite structures [9,10]. During the joining process, thermoplastic rivets are heated above the melting temperature, which allows the creation of a squeeze flow of the rivet material into the internal composite structure. However, thermoplastic welding methods, such as ultrasonic welding, offer several advantages over the above-mentioned joining techniques, such as short process times and strong dependable joints without significant surface preparation efforts [8,11–15]. Ultrasonic welding is one of the most commonly used welding methods besides resistance welding [11,13,14] or conduction welding [16] for joining thermoplastic structures. It uses ultrasonic energy at high frequencies (20–40 kHz) to produce low peak-to-peak amplitude (50 to 100 μm) mechanical vibrations. In ultrasonic welding, the heat required to melt the thermoplastics is generated locally. This occurs through friction between the joining constituents and viscoelastic damping [17,18]. Ongoing research has focused on developing ultrasonic welding solutions to replace gluing and riveting for primary load-carrying aerospace structures [13,18,19]. These applications are generally larger structures based on reinforced high-performance thermoplastic materials, such as polyetheretherketone (PEEK), polyphenylene sulphide (PPS), and polyetherimide (PEI) [13,20,21]. Furthermore, continuous ultrasonic welding systems that are capable of continuously joining large flat and double-curved structures have been shown to be a fast and feasible welding technique [20,22].

The MM-Welding® LiteWWeight® technology is especially designed for sandwich structures with honeycomb and foamed core materials (Figure 1). Ultrasonic energy is used to insert a thermoplastic connection element (pin) through the top layer into the porous structure of the core. Owing to the ultrasonic excitation, the thermoplastic connection element melts at the interface, infiltrates the porous structure, and creates a highly interlocked bond with the substrate material. The goal of this study is to adapt the knowledge-based approach for modelling the LiteWWeight® process targeting interior automotive applications e.g., fixation elements on casing panels in the trunk.

The MM-Welding® LiteWWeight® process is divided into four phases:

- Phase 1: Ultrasonic device moves to target position to start joining process.
- Phase 2: Pressure is increased on the thermoplastic pin until the trigger force is reached that starts the ultrasonic excitation. The pressure continues to increase until the pin dynamically pierces the upper face sheet resulting in a local face sheet collapse.
- Phase 3: The thermoplastic pin moves through the substrate. Tolunay et al. [23] suggested that heating is solely due to mechanical dissipation of work through viscoelastic deformation. In the case of a flat energy director, the heating mechanism, as suggested by Zhang et al. [24] and Villegas et al. [17,18] also consists of frictional dissipation at the interface. The heat sources are based on two effects: bulk phenomena and friction-based interfacial heat sources. Based on available literature [23,25], heat generation appears in the material itself from the viscoelastic dissipation (bulk phenomena). This is based on the total or partial contact of the sonotrode (hammering effect), the loss modulus of the material, and the frequency and amplitude of the imposed strain.

- Phase 4: Intimate contact: During the insertion, the contact between the parent materials and connection element evolves as the surface of the connection element melts and flows/squeezed into the porous material. To take into account this effect Levy et al. [25] proposed an adapted model from Lee et al. [26] dependent on a dimensionless scalar definition that allows an assessment of the process quality.

This study aims to assess whether it is feasible to create a dynamic numerical impact model of the LiteWWeight® process that allows an evaluation of the connection quality. The impact model calculates the stresses, strains, and damage rates for pin and sandwich structures. Additionally, the numerical model enables the calculation of heat from friction and plastic deformation. The model was verified and validated based on a comparison with experimental tests. The data provided by the ultrasonic device, such as force, displacement, and ultrasonic amplitude, depending on time, were used for the reconstruction of the sonotrode displacement boundary condition (BC). Computed tomography (CT) scans were used to reconstruct the sandwich geometry and pin insertion position. Furthermore, a high-speed camera was used to visually assess the LiteWWeight® process and calculate the hammering coefficient. Additional studies were performed to evaluate the damage rate and the temperature distribution of the pin during joining.

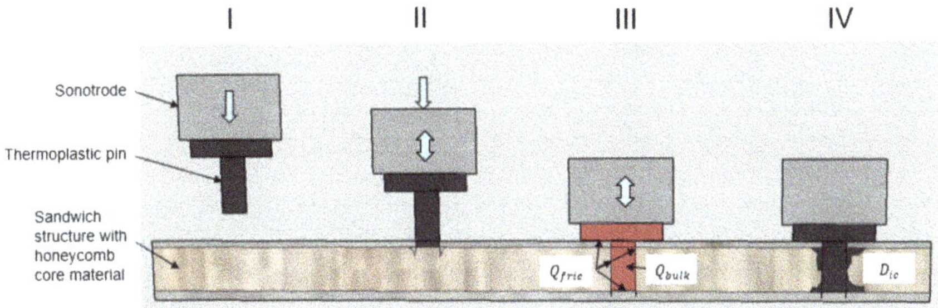

Process stage	Pin´s movement to the welding position	Collapse of the top facesheet	Partial insertion of the pin into the sandwich structure	Reaching final pin position, interlocking and cooling
Details	Movement continues until pin touches the sandwich and specified reaction force is reached	Specified pressure is applied while ultrasonic excitation is active	Heating by friction and plastic deformations arises on pin´s surface and surrounding foam material	Material flows into the "porous" structure to form an interlock with the substrate

Figure 1. Schematic overview of the MM-Welding® LiteWWeight® process. Adapted from Ref. [15].

2. Materials and Methods

2.1. Materials

Thermoplastic material: The thermoplastic pins were made from the fibre-reinforced polymer PA66 GF30 (BASF, Ultramid A3WG6, Ludwigshafen, Germany) and were produced by injection moulding. The use of a glass fibre-reinforced polymer is advantageous because of the increased Young's modulus and mechanical loss factor of the material. Therefore, the ultrasonic vibrations were converted into heat more efficiently, which was crucial for the MM-Welding® LiteWWeight® process. The thermoplastic pins were dried at 90 °C for 8 h because of the influence of the absorbed humidity in PA66 GF30 [15]. The pins resulting elastic modulus is 5500 MPa at room temperature. Thus, a temperature-dependent elastic modulus was implemented into the numerical model.

Sandwich-structured composite: A lightweight panel, which is typically integrated in the automotive sector as an interior component, was used. It is based on face sheets made of glass fibre-reinforced polymers and a honeycomb core structure made of cardboard. The components were bonded together using a polyurethane resin that partially expanded into the core layer. The elastic modulus for the face sheet is 4550 MPa, 400 MPa for the honeycomb cardboard internal structure and 300 MPa for the foam. The overall thickness of the panel was 16 mm, whereby top and bottom face sheets were 0.5 mm. The length and width of a honeycomb element was respectively 13 mm and 6 mm.

2.2. Material Characterisation

Various methods have been used to characterise the mechanical and thermal material properties of the pin material PA66 GF30 and sandwich material to identify parameters for the finite-element simulation model. First, differential scanning calorimetry experiments were performed to analyse the crystallisation behaviour at different cooling rates and to derive heat capacities. Temperature- and frequency-dependent dynamic mechanical analysis measurements were performed for the purpose of storage and loss modulus. Furthermore, the following mechanical tests were conducted: tensile test on the face sheet substrate material, compression behaviour of the pins, and the sandwich substrate. The face sheet tensile test is essential for the implementation of the Hashin damage criteria. A detailed representation of the data is documented in Brauner et al. [15] and summarised in Appendix A (Table A1).

2.3. Ultrasonic Equipment and Process Monitoring Tools

In this study, a servo-driven ultrasonic welder (Dukane iQ Servo, Prague, Czech Republic) was used. The equipment monitored the force, displacement, amplitude, frequency, velocity, and energy with a sampling rate of 1000 Hz. The ultrasonic insertion of LiteWWeight® pins was optimised in a preliminary study by systematic variation of relevant processing parameters using a design of experiments [15]. Hence, optimised reference process parameters were used in this study. A velocity of 45 mm s^{-1} and an amplitude of 76 µm operating at 20 kHz were applied.

Because the LiteWWeight® process is highly dynamic, high time resolution methods were exploited to allow further data analysis and interpretation. A FASTCAM SA5 high-speed camera (Photron, Tokyo, Japan) was used to analyse effects such as pin movement, hammering (contact between the pin and sonotrode), and potential pin-tip failure. To derive the concept of the hammering coefficient, which plays an important role in calculating heat by friction and dissipation, a MATLAB algorithm was developed and coupled with a high-speed camera to track various points on the pin and sonotrode through the process (Figure 2). High-speed videos were acquired at 75 kHz, which was the limit due to the imaging resolution and the available light sources. An ideal sinus wave at 20 kHz was considered during this process. Therefore, 3.75 points per cycle were obtained by high-speed measurements to characterise the sinusoidal wave. The Kanade–Lucas–Tomasi (KLT) algorithm implemented in MATLAB (Version 9.8, V Natick, MA, USA) Computer Vision Toolbox was used to track various points of the sonotrode and pin [27]. The algorithm returned the trajectories of positions that were successfully tracked throughout the process. The most suitable trajectory for both the sonotrode and pin was extracted depending on the quality and stability of the signal.

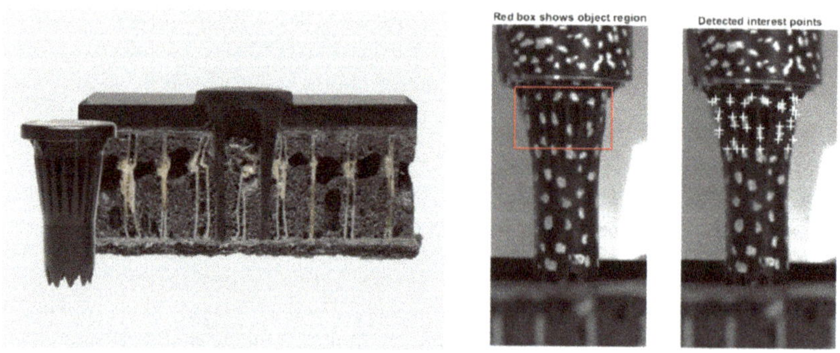

(a) (b)

Figure 2. (**a**) Thermoplastic pin inserted in lightweight panel for high-strength fixation; (**b**) Detected points of interested on thermoplastic pin by usage of the Kanade–Lucas–Tomasi point tracking algorithm.

Additionally, the load cell KM40d (ME-Meßsysteme GmbH, Hennigsdorf, Germany) was used to record the reaction force below the sandwich panel with a high time resolution of 1 kHz. The unfiltered force curve results from oscillatory reaction forces combined with measurement errors due to high sampling rate. The measured force signal was smoothed using a Gaussian filter method (see Figure 3), which showed reasonable agreement with the exported machine data. Equivalent post-processing was applied to the reaction forces derived by the numerical model.

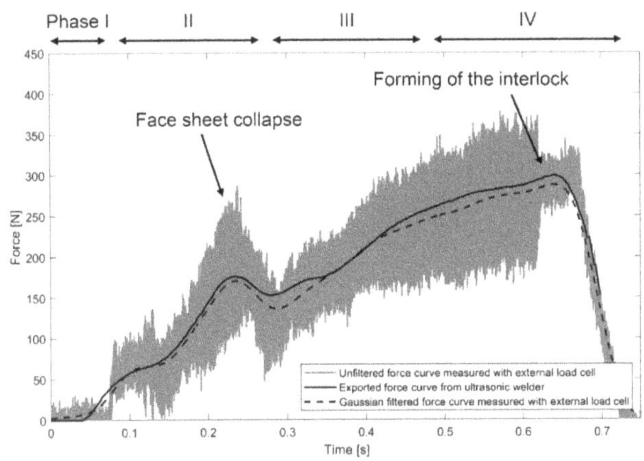

Figure 3. Comparison of force curves derived from different sensor data.

Additionally, CT scans were performed. The scans presented an inside view of the inserted pin, highlighted the mechanical strength of the pin and the surrounding foam content. Furthermore, it allowed quick measurement of the densities of the components used and showed possible weaknesses that were introduced during manufacturing (e.g., porosities within the pin due to injection moulding). The mechanical strength of the pin was measured by a pulling out procedure in a Zwick 100 kN universal tensile test machine (Zwick, Ulm, Germany) at a testing speed of 5 mm/min to determine the maximum force. Furthermore, a partial insertion study was performed with variations in insertion collapse length. This method allowed the assessment of the damage rate of the thermoplastic pin tips at various phases during the process.

2.4. *Numerical Methods*

The commercial software Abaqus (Version 6.13, Providence, RI, USA) was used for the simulation of the heat transfer, elastic and plastic deformations, and failure in pin and substrate elements. A fully explicit coupled thermal-mechanical dynamic analysis was implemented. This full coupling was needed because most of the heat arose from friction and dissipation, and the material properties are temperature-dependent.

2.4.1. Thermo-Mechanical Model

Three general heat sources arose in the pin and substrate during the process: heat from friction, plastic deformation, and ultrasonic energy dissipation. The heat fluxes generated on the contacting surfaces of the pin and substrate were estimated according to the following:

$$q_p = q_k - f_1 q_g, \tag{1}$$

$$q_s = -q_k - f_2 q_g. \tag{2}$$

Here, q_p and q_s are the heat fluxes generated on the pin and substrate contacting surfaces, respectively, q_k is the heat flux due to conduction, and q_g is the heat flux from frictional energy dissipation. $f_1 = f_2 = 0.5$ are the fractions of evenly distributed heat generated on the first and second surfaces, respectively [28]. The values for f_1 and f_2 were set as equal values, because both joining partners are polymer materials and therefore, the thermal properties like heat capacity, density, and thermal conductivity are similar. The heat generated by friction in Abaqus/Explicit was calculated using the factor η_{fr}, which defines the fraction of frictional work converted to heat. The heat flux generated by the frictional heat generation is given by

$$q_g = \eta_{fr} \tau \Delta s / \Delta t, \tag{3}$$

where τ is the frictional stress, Δs is the incremental slip, and Δt is the incremental time. The heat flux due to conduction is assumed to be of the form

$$q_k = \kappa \Delta \theta, \tag{4}$$

where κ is the heat-transfer coefficient, and $\Delta\theta$ is the temperature difference between the two sides. In [28] it was shown that heat from energy dissipation appeared in the domain as a thermal source according to the following equation:

$$\dot{Q}_{bulk} = \frac{\alpha^2 \omega E_{loss} \varepsilon : \varepsilon}{2}, \tag{5}$$

where $\dot{Q}_{bulk}$ is the heat rate, α is the hammering coefficient, ω is the frequency, E_{loss} is the elastic loss modulus, and ε is the strain tensor. According to the modelling and experimental results presented in [18,23], the heat from friction on the interface of joined parts was significantly higher than heat from energy dissipation. Therefore, heat from ultrasonic energy dissipation was neglected in the Abaqus model but was calculated in the next step using the evaluated values of hammering coefficient and strains depending on time.

Abaqus allows the calculation of the inelastic heat fraction, which defines the amount of heat generated by mechanical dissipation associated with plastic strain. This term can be introduced as a source of coupling for the thermal-mechanical analysis. The inelastic heat fraction plays a significant role in LiteWWeight® technology, because extensive inelastic deformations occur rapidly in a material whose mechanical properties are temperature-dependent, and heat has no significant time to dissipate. The heat flux per unit volume was estimated according to the equation

$$r^{pl} = \eta_{pl} \sigma : \dot{\varepsilon}^{pl}. \tag{6}$$

Here, η_{pl} is a user-defined parameter, σ is the stress, and $\dot{\varepsilon}^{pl}$ is the rate of plastic straining. General contact was defined in Abaqus/Explicit for the simulation of contact and interaction problems. All surfaces were automatically defined. This default surface contained all exterior element faces, all analytical rigid surfaces, and all edges in the model, as well as the nodes attached to these faces and edges. The general contact algorithm activated and deactivated the contact faces and contact edges in the contact domain based on the failure status of the underlying elements.

The Coulomb friction model was implemented to carry shear stresses by the definition of critical shear stress, at which sliding of the surfaces started as a fraction μ_{fr} of the contact pressure.

$$\tau_{crit} = \mu_{fr} p, \tag{7}$$

where μ_{fr} is the same in all directions. Contact pressure-overclosure relationships are defined as a linear function of the clearance between the surfaces. In a linear pressure-overclosure relationship, the surfaces transmit contact pressure when the overclosure between them, measured in the contact (normal) direction, is greater than zero.

Abaqus/Explicit predicts material progressive damage and failure based on the undamaged elastic-plastic response of the material, damage initiation criterion, and damage.

Plastic stress-strain ratio is defined from the experiments for the pin, and plastic properties of honeycomb cells and foam are known as a range of possible values (plastic and damage properties of internal substrate structure is a subject of model calibration procedure). In both cases, the von Mises yield surface was used to define isotropic yielding. It is defined as the value of the uniaxial yield stress as a function of the uniaxial equivalent plastic strain.

The pin destroyed the upper face sheet and the internal structure of the sandwich panel through the substrate. Therefore, failure models and element removal criteria were defined and integrated into the model. Further governing equations related to Abaqus/Explicit were complemented as supplemental material.

2.4.2. Assessment of the Connection Quality

The main goal of the LiteWWeight® technology is to form an interlock between the pin and substrate and provide a high value of pullout force. The form-locking process is a function of the process parameters (pressure, temperature, and time) and the geometry of the contacting surfaces. A geometric description of the surfaces is critical for modelling. Surface topography can be described as a nonstationary random process [29] or fractal distribution [30]. With simplified approaches, the surface profile can be interpreted as a series of periodic sine or cosine functions with different amplitudes and frequencies. A model of intimate contact by Dara and Loos [31] represents the surface as a set of rectangles of different sizes. Lee and Springer [26] and Mantell and Springer [32] proposed models with simplified geometry represented as a series of rectangles of the same size. In this study, a simplified 'squeeze flow' model by Mantell and Springer [32] was implemented as

$$\dot{D}_{ic} = \left(\frac{D_{ic,0}}{D_{ic}}\right)^4 \left(\frac{a_0}{b_0}\right)^2 \frac{p}{\mu(T(t))}. \quad (8)$$

Here, D_{ic} is the degree of interlock, $D_{ic,0}$ is its initial value, $\mu(T(t))$ is the viscosity of the pin material in a liquid state, p is the pressure, and a_0/b_0 is the ratio derived from the surface roughness.

The calculation of the degree of interlock according to expression (8) for the LiteWWeight® technology is not a trivial task because the interface surface is non-constant, elements of pin and substrate were deleted according to the defined element removal criteria, and the whole pin moves relative to the substrate. For the validation of the proposed approach for the performance quality prediction, five different insertion setups were considered: reference, faster insertion speed, lower insertion speed, higher US amplitude, and lower US amplitude. Three trials were performed per setup. The specifications are presented in Table 1.

Table 1. Ultrasonic process setup for assessment of the connection quality.

Setup	Insertion Speed [mm/s]	Amplitude [µm]
Reference	45	76
Slow	40	76
Fast	50	76
High Amplitude	45	92
Low Amplitude	45	62

An Anton-Paar MCR 300 rheometer was used to perform rheological measurements that were used to describe the temperature-dependent viscosity of PA66 GF30. A plate-plate configuration was used for the measurements. The plate diameter was 25 mm, with a gap thickness of approximately 1 mm. The samples were dried before testing for 12 h at 90 °C. Additionally, the samples were exposed to nitrogen during the measurement because of the accelerated thermal degradation of PA66 GF30 at high temperatures. Furthermore, the confocal laser microscope (Keyence VK-X1100) enabled the characterisation of the surface roughness at different locations on the pin surface. The surface roughness values were evaluated with one-dimensional line roughness measurements according to standard DIN EN ISO 4287.

3. Results and Experimental Validation

The overall objective of this section is the validation of the numerical model by comparison with experimental results. The numerical model should adequately predict process performance which was represented by the degree of interlock that is dependent on the input parameters temperature, stress, and failure fields. After the insertion process, the most important process signals were exported from the ultrasonic welder, as shown in Figure 4. The force curve clearly presents phase 2 (first force peak, collapse of face sheet) and phase 4 (second force peak, forming the interlock) as described in Figure 1.

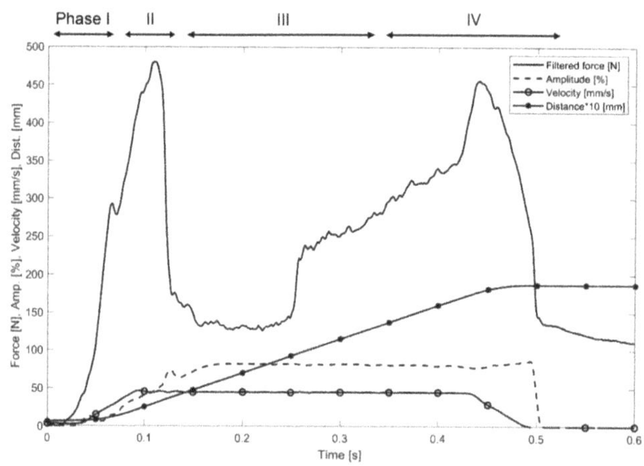

Figure 4. Process signals such as filtered force, amplitude, and velocity during the LiteWWeight® joining process of a thermoplastic pin with the sandwich-structured composite panel.

3.1. Geometry Reconstruction

CT scans were analysed before and after the insertion process at the same position for validation. Several cross-sections of the sandwich structure were imaged to replicate an approximated structure in the impact model. Figure 5 presents a detailed view of the sandwich structure. It was possible to examine the amount of foam, the porosity of the foam, and its position on the cell structure. The insertion position in the substrate is highlighted with yellow coordinate lines to create a three-dimensional (3D) impression from different views (front view, top view, and side view). The geometry of the sandwich structure was reconstructed in detail because of the high variability of the internal sandwich structure and its strong influence on the process outcome. For this purpose, CT scans of the substrate sample were cut into eight slices in the thickness direction, and the sandwich geometry was represented as a layered structure with seven levels. At every level, the honeycomb cell walls were presented in a pattern corresponding to the sandwich structure in the insertion position (Figure 5).

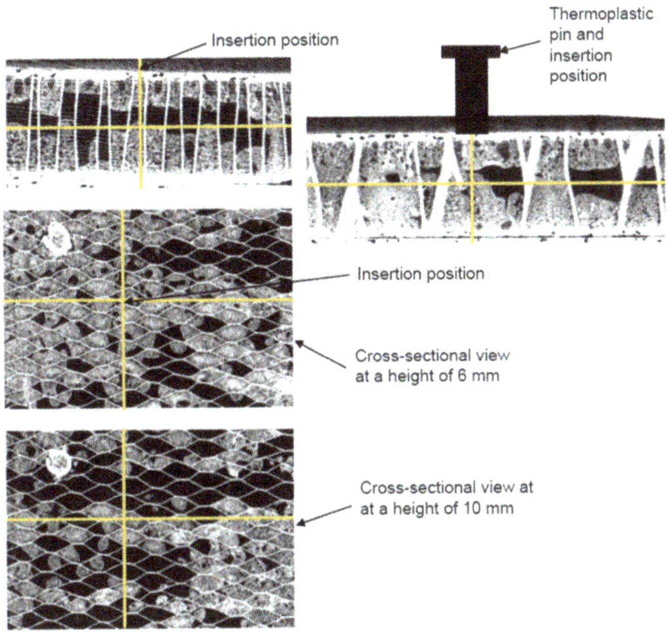

Figure 5. Computer tomography (CT) scanned sandwich structure with corresponding insertion position highlighted with yellow coordinate lines.

The wall thickness (0.5 mm) was measured on the basis of CT scans. The internal space in each honeycomb cell at every level was filled with a 3D hexagonal prism domain, depending on the foam volume content in this region. The internal space of the cell was filled if the CT scan analyses showed that foam occupied more than 50% of the 3D hexagonal prism volume. Performing simplification of the foam distribution to FEM geometry allows the consideration of relatively large void areas inside the honeycomb, neglecting small local variations. The top and bottom composite surfaces were presented as two-dimensional shell structures defined as the skin of the 3D sandwich domain. The proposed method, for the simplified representation of a real 3D lightweight structure, considers non-uniform foam distribution inside the honeycomb with the composite face sheets, and its block-wise domain allows sweeping of the triangular mesh on the face sheets through the thickness direction, both of which provide an increased calculation accuracy in the impact region and reduce computational time.

The model assembly consists of three parts: ultrasonic sonotrode, thermoplastic pin, and sandwich structure (Figure 6). The sonotrode material is titanium alloy (Ti-6Al-4V) and is therefore significantly stiffer than the pin. Accordingly, the sonotrode is considered to be a rigid surface with a shape corresponding to the sonotrode used in the experiment. The composite face sheets were defined as skins that are bonded to the surface of the three-dimensional sandwich domain via a tie constraint with consistent mesh. Skin reinforcements influence contact calculations through the change of contact penalty stiffness depending on the skin thickness. General contact in ABAQUS/Explicit was defined for all the external surfaces. General contact enforces contact constraints using a penalty contact method, which searches for node-into-face penetrations. For node-to-face contact, forces depend on the penetration distance. They were applied to the slave nodes to oppose the penetration, while equal and opposite forces act on the master surface at the penetration point.

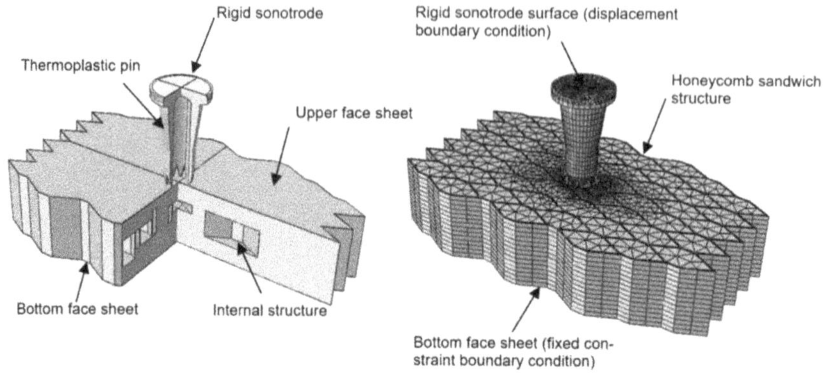

Figure 6. Overview of model assembly (thermoplastic pin, face sheet and internal structure) and resulting mesh with two node-based boundary conditions.

The pins domain was meshed using 12′647 linear coupled temperature-displacement tetrahedron elements. The substrate's domain was meshed using 62′656 linear coupled temperature-displacement triangular prism elements and composite shells were meshed with 3′194 triangular elements. Both sandwich and pin domains were meshed using sweep meshing routine, face sheets were meshed with free triangular mesh. The average volume of the pin's elements is 0.95 mm^3, of the substrate's elements is 1.41 mm^3 and the average area of the face sheet elements is 1.24 mm^2. Abaqus/Explicit automatically evaluates the time step on the base of the minimal element size and mechanical properties of the considered domains. Therefore, proposed finite-element mesh is a result of the compromise between the simulation accuracy and total calculation time. Defined finite-element mesh provides an average time step being equal to 2×10^{-8} s, which leads to approximately 25 million iterations for the simulation of 0.5 s of the joining process. The numerical model was solved using 16 processors within 27 h. Decrease of the average element size leads to the significant increase of the calculation time, while increase of element size leads to no solver convergence due to the hour-glassing problem.

3.2. Boundary Condition Reconstruction

The displacement and its derivatives were set to zero and the temperature was set to 20 °C for all domains at the beginning of the simulation. The model contained two node-based boundary conditions: displacement for the rigid sonotrode and fixed constraint (zero displacements and rotations) for the sandwich bottom face sheet (see Figure 6). The vertical displacement of the rigid sonotrode was defined based on the experimental data. It takes into account the linear displacement of the sonotrode, oscillation frequency, and amplitude according to the expression

$$u_{horn} = u_{web} + A_{web} sin(2\pi\omega t). \tag{9}$$

Here, u_{web} is the vertical displacement of the sonotrode measured by the ultrasonic system, A_{web} is the time-dependent value of the oscillation amplitude, and ω is the oscillation frequency. Displacements in other directions, as well as all the rotations were set to zero.

3.3. Stress and Temperature Distribution

Figure 7 presents an overview of the resulting thermal-mechanical numerical model. The material properties are defined according to the experimental studies and are presented in Appendix A (Table A1). Figure 7a shows the stress distribution in the pin and substrate. The maximum stresses arise in the pin tips at the end of phase II (maximum stress time corresponds to the first reaction force peak time, when the collapse of the face sheet occurs, see Figure 4) and phase IV (time corresponds to the second reaction force peak, pin reaches the bottom of the substrate, see Figure 4).

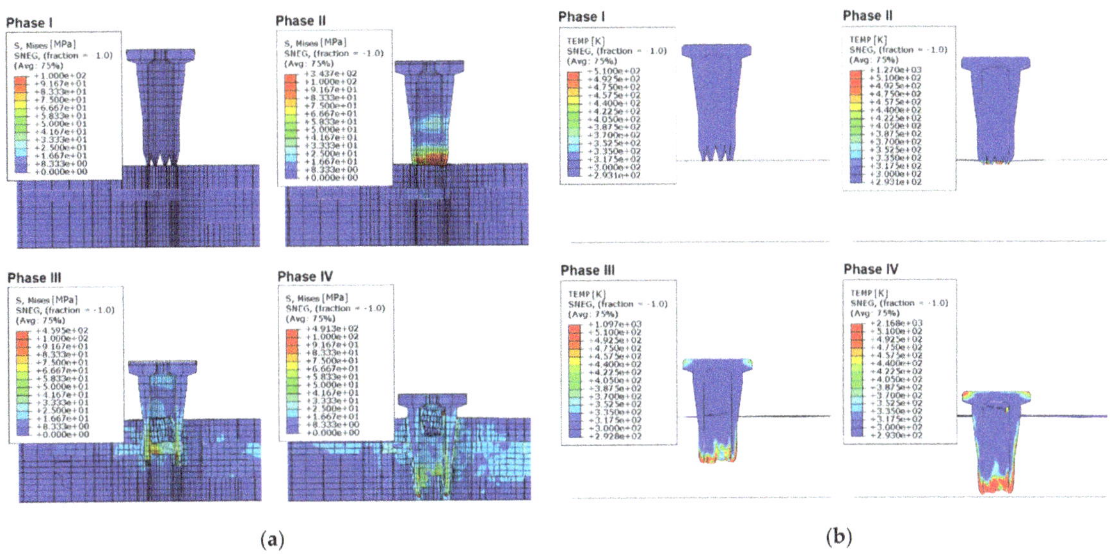

Figure 7. Simulation of the MM-Welding® LiteWWeight® process: (**a**) stress distribution within the four process phases; (**b**) Temperature distribution within the four process phases.

The solution of the heat-transfer equations is shown in Figure 7b. Most of the pins plastic deformations and friction arise in the tip area owing to the permanent contact with the substrate during the pins' movement, which leads to a temperature increase in that region. Heating of the head of the pin is explained by the friction between the head and sonotrode internal walls. The simulated temperature distribution corresponds to the melted areas observed after the process. A detailed comparison between the measured and simulated temperatures is provided later on.

3.4. Reaction Forces

The reaction force curve represents the main phases of the process (Figure 4). Therefore, the proposed numerical model should provide the same values of the reaction force as measured for the corresponding process phase. The sampling rate for the reaction force evaluation in the simulation was equal to 1/10 of the oscillation period. The reaction force signal provided by the ultrasonic welder was smoothed by a Gaussian filter, as described in Figure 3; for the comparison of simulated and measured data, the simulated reaction force curve was, therefore, smoothed by an identical filter. A comparison of the measured and simulated smoothed force values with time is presented in Figure 8. The calculated reaction force provided a reasonable correlation with the measured reaction force both qualitatively and quantitatively. The oscillations within the numerically modelled force signal arose from the element deletion criteria.

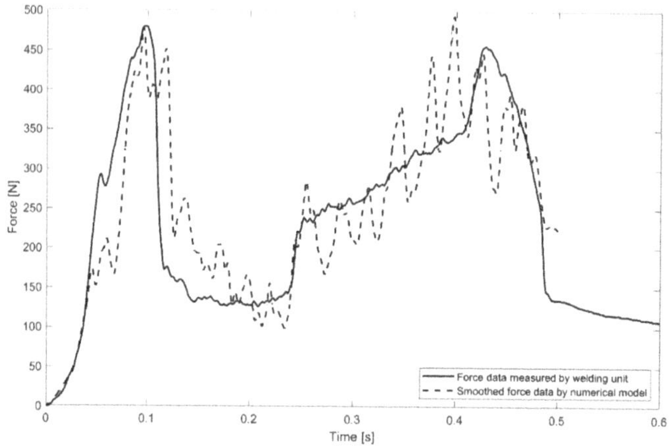

Figure 8. Numerically modelled and experimentally measured reaction force of the sonotrode.

3.5. Damage Rate

A 'partial insertion' study was performed to measure the destruction rate of the pins tips on different phases of the process. The destruction rate in the final phase of the process plays a significant role in the formation of an interlock between the pin and bottom composite face sheet. A high destruction rate leads to the shortening of the pins body, and possibly to the situation where the pin does not reach the bottom face sheet and does not form an interlock, thereby significantly reducing the joint quality. On the other hand, a low destruction rate could lead to penetration of the bottom face sheet by the tips, which is unacceptable in terms of the visual appearance.

In the frame of the 'partial insertion' study, various collapse lengths were set in a series of trials. After partial insertion, the pin was pulled out and the reduction in the tip height was measured. Owing to the strong influence of the internal structure on the process, the pin lost different percentages of the original tip's height with the same collapse distance at different insertion locations. However, in most cases, the pin destruction rate increased significantly after the upper face sheet collapsed and after reaching the bottom face sheet. In general, the pin-tip damage rate evolution can be described in four steps:

- Before the upper face sheet collapse (collapse length is 0–1.5 mm): tips were partially melted and slightly deformed; approximate loss of original tip's height was 10%.
- After the upper face sheet collapsed (collapse length is 1.5–3 mm): tips were partially destroyed; approximate loss of original tip's height was 10–30% depending on local stiffness under a single tip.
- Before the pins reached the bottom face sheet (collapse length is 3–16 mm), the destruction rate was slightly increased. Since the substrate (foam) material was much softer than the pins, the pin did not meet significant resistance during its insertion through the core of the sandwich structure. However, during this step, the temperature of the tips increased significantly owing to friction.
- After the pins reached the bottom face sheet (collapse length is 18 mm): the tips lost most of their original height.

A comparison between the results of the 'partial insertion' study and simulation for the key steps of the joining process is presented in Figure 9. The correlation of the numerical model and experimental trials, by verifying the damage behaviour during the insertion process, showed a reliable quantitative result. The simulation of the tip destruction during the joining process allowed the optimisation of the pin design to provide the maximum interlocking surface without penetration of the bottom face sheet. The optimisation of the pins design based on the numerical models is the subject of future studies.

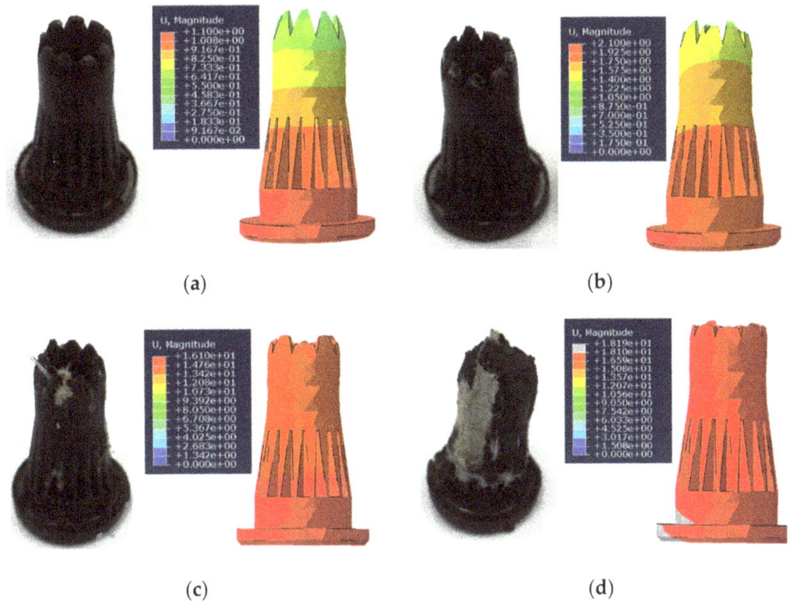

Figure 9. Observed and simulated pin damage rate at various collapse lengths: (**a**) Collapse length of 1 mm (before face sheet collapse); (**b**) Collapse length of 2 mm (after face sheet collapse); (**c**) Collapse length of 16 mm; (**d**) Maximum collapse length of 18 mm.

3.6. Hammering Coefficient

The idea of calculating the hammering coefficient is based on the determination of the contact and non-contact points within the measured signals. With the labelled points, the proportion can be calculated and referenced as the hammering coefficient. First, both data arrays (sonotrode and pin) were normalised by polynomial regression. Subsequently, the signal for clean oscillations was evaluated. The next step was to sum up the distances between the pin and the sonotrode and divide it by an averaged value. Then, the 'contact' and 'non-contact' points were defined and the occurrence was calculated. The hammering coefficient was evaluated by dividing both the sums. To date, this approach is the most robust and repeatable method.

Evaluation of the hammering coefficient is an important step in understanding sonotrode-pin interactions during the process. For the experiment, the hammering coefficient was evaluated using the MATLAB tracking algorithm, as described in the experimental methods. The hammering coefficient was also evaluated on the basis of the simulated reaction force. When the reaction force is equal to 0, there is no contact between the pin and sonotrode. The hammering coefficient value within the current time range can be evaluated as the ratio of non-zero force values to the total amount of force values. A comparison between the measured and simulated hammering coefficients is shown in Figure 10. The MATLAB tracking algorithm is not able to measure the hammering coefficient for the whole process, since no oscillations were existing before 0.05 s, and most of the tracking points were covered by the substrate and were not observable with the high-speed camera after 0.25 s.

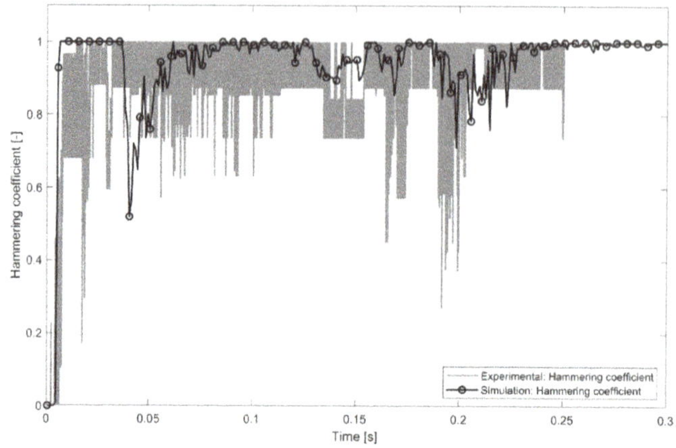

Figure 10. Hammering coefficient calculated on base of MATLAB point tracker and FEM.

Both FEM and MATLAB tracking algorithms show that after the ultrasonic vibrations started, the hammering coefficient was close to 1 (full contact) with small local reductions. The local hammering coefficient predicted by the FEM and MATLAB tracking algorithms did not always correlate with each other. This could be explained by the limited accuracy of the implemented methods (coarse finite-element mesh or low resolution of the high-speed camera) or by the inaccuracy of the substrate geometry reconstruction since local variation in the substrate's foam content plays a significant role in the pins' movement.

3.7. Temperature Distribution

Abaqus allows the consideration of heat from plastic deformation and friction. The Abaqus results for temperature are presented in Figure 11. In the current study, no direct temperature measurements were possible based on the fast process. Temperature measurements were performed immediately after insertion (the pin was pulled out shortly after the insertion and temperature was measured in several pin regions) for the comparison of the temperature distribution in the experiment and simulation. The difference between the experiment and simulation could be explained by the time elapsed from the insertion end to the measurement. During this time, temperature redistribution within the pin domain occurred and free convection with the room environment cooled the pin. However, spatial distribution of the temperature in the simulation was close to that in the experiment, and the pins head in experiment had signs of melting, which meant that the temperature there was higher than 240 °C during the joining.

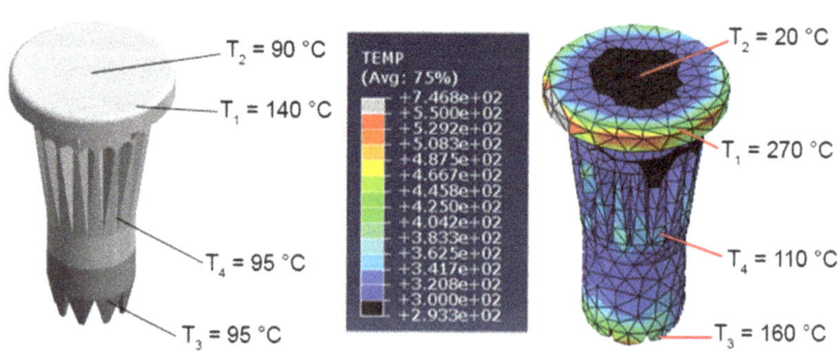

Figure 11. Temperature measurement after the insertion and pulling out and simulated temperature after the insertion.

The temperature increase due to the elastic deformations was calculated using Equation (5) for the heat from the ultrasonic energy dissipation. According to this equation, the temperature grew only by about 0.1 degree after 0.5 s, which was significantly lower than the heat from the plastic deformations and friction and was, therefore, neglected in further calculations of the degree of interlock.

3.8. Connection Quality Evaluation

Figure 12a shows a one-dimensional representation of the surface roughness of the pins, resulting in $R_z = 6.44$ μm and $R_a = 0.93$ μm. The measurements were used to calculate the degree of interlock. Therefore, the approximated proportion $a_0/b_0 = 0.3$ was extracted for use in Equation (8). Furthermore, Figure 12b presents the frequency- and temperature-dependent viscosity measurements of the PA66 GF30 pin material.

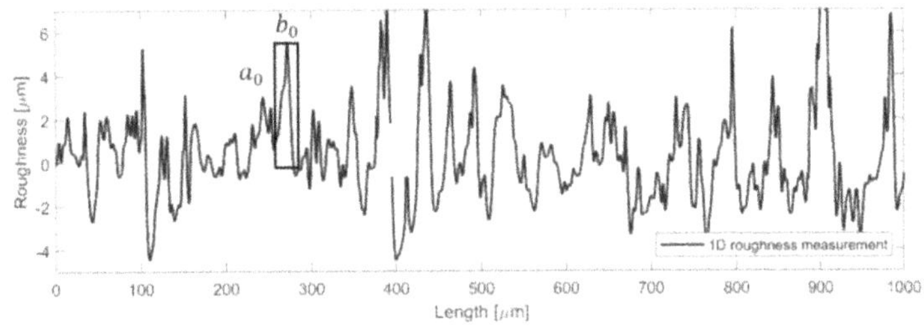

(a)

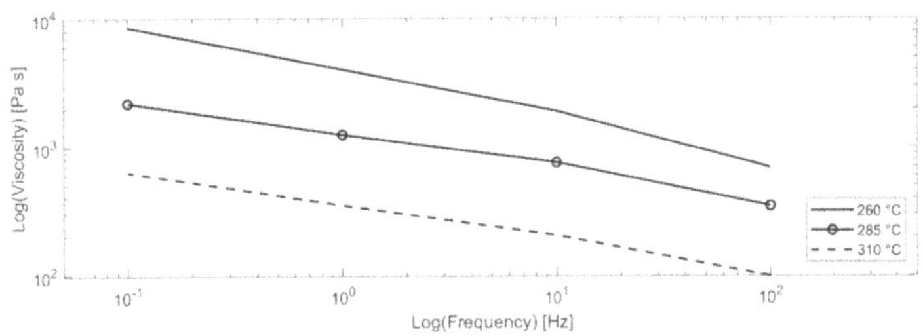

(b)

Figure 12. (**a**) One–dimensional surface roughness measurement on the 'connecting' surface of the pin; (**b**) Frequency- and temperature-dependent viscosity measurement of the pins material PA66 GF30.

The sonotrode boundary conditions were reconstructed for every setup on the basis of the exported data according to Equation (9). Since Abaqus software does not support the calculation of the connection quality, a MATLAB routine was developed for the evaluation of the degree of interlock on the base of thermal and stress distribution exported from the finite-element model. Thus, the Abaqus solution for the temperature and pressure was exported for every finite element (except the head region and elements that were destroyed during the process) for every time step starting from the Phase III (after pins vertical movement slows down), and implemented into Equation (8) for the degree of interlock. According to the simulated and experimentally-observed temperature distributions, only the tip and energy director area of the pin were subjected to melting. Therefore, only a small percentage of all nodes reached a fully interlocked state ($D_{ic} = 1$). The degree of interlock distribution among the defined nodes is presented in Figure 13 with a smoothed probability density function.

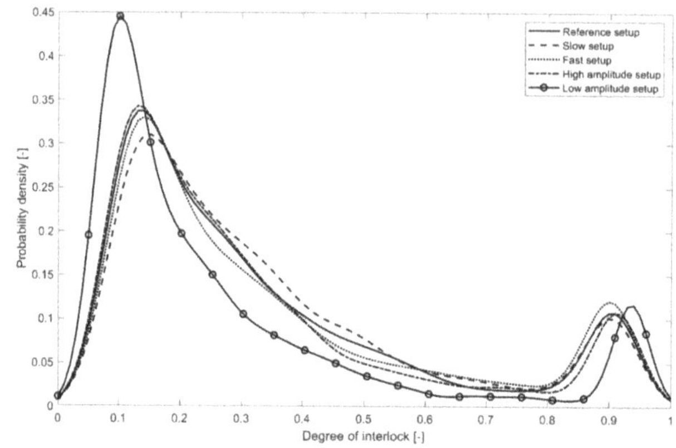

Figure 13. Degree of interlock distribution among nodes for the performed setups.

The calculated average degrees of the interlock and measured pullout forces are presented in Table 2. Experimental trials show that the process did not depend significantly on the small variations in the parameter setup. Only the 'Low Amplitude' configuration demonstrated significant (10%) reduction of the average pullout force, while other setups demonstrated small (1–3%) variation of the pullout force in comparison with the 'Reference' setup. The developed numerical model predicted 16% reduction (vs. 10% for the experimental trials) of the average pullout force for the 'Low Amplitude' setup, which was considered a reasonable agreement with the experiment. Nevertheless, the numerical model was not able to predict a small (3%) reduction in the average pullout force for the 'Fast' setup. It also predicted a small (4%) reduction in the average pullout force for the 'High Amplitude' setup, while the experiments showed a small increase (1%). The developed numerical model predicted only a significant change in the average pullout force depending on the process setup. Therefore, the proposed numerical approach can be used to define the process setup in the case of new pin or substrate designs, where significantly different setups must be tested, although the model was not able to track the changes in the final joint quality within the small variation of the parameter setup. Thus, a prediction accuracy between 94–99% depending on the setup was achieved.

Table 2. Measured pullout forces and corresponding simulated average degree of interlock (DIC).

Setup	Average Pullout Force (% to Reference Value)	Average Degree of Interlock (% to Reference Value)
Reference	530.2 N (100%)	0.3762 (100%)
Slow	525.83 N (99%)	0.3824 (101%)
Fast	511.76 N (97%)	0.3790 (101%)
High Amplitude	534.43 N (101%)	0.3600 (96%)
Low Amplitude	477.8 N (90%)	0.3166 (84%)

4. Conclusions

In this study, the MM-Welding® LiteWWeight® technology was presented, and a thermo-mechanical numerical model that predicts the joining quality was developed. Based on the results obtained, the following conclusions can be highlighted:

- Material properties of the pin and sandwich were defined using various experimental measuring techniques.
- The geometry in the numerical model was reconstructed based on computed tomography scans to ensure identical process conditions, pin position, and local foam distribution.
- The numerical model successfully calculated the damage rates and heat from friction and plastic deformations of the pin and sandwich structures.

- The model was verified and validated based on comparative experiments.
- The routine for the prediction of the connection quality from the finite-element model temperature and stress distribution was derived based on the degree of the interlock approach, which only showed reasonable agreement for significantly different process setups resulting in a prediction accuracy between 94–99%. Despite this fact, the methodology applied suits the potential for lowering the effort towards new pin and sonotrode materials.
- The numerical simulation and data analyses conducted during the process enabled the discrimination and identification of critical process and design features, such as hammering, damage rate or reaction force.

Presented results will not only accelerate process development cycles, but also provide the potential for an advanced online process quality monitoring system. Predictive capabilities of the developed model enable the possibility of numerical-based processes and pin geometry optimization, which is the focus of the future studies.

Supplementary Materials: The following are available online at https://www.mdpi.com/1996-1944/14/20/6005/s1, Equations (S1)–(S24): Governing equations.

Author Contributions: Investigation, methodology, validation, writing—original draft preparation, visualization, L.Z. Methodology, formal analysis, investigation, writing—original draft preparation, I.Z. Supervision, project administration and funding acquisition, C.B. Project administration and funding acquisition, M.R. Supervision, G.E. Formal analysis, V.B. Formal analysis, S.G. Project administration, R.P. All authors have read and agreed to the published version of the manuscript.

Funding: This study was supported by the national funded project LiteWWeight connection for sandwich panels [grant number 27066.1], which was supported by Innosuisse, Switzerland.

Institutional Review Board Statement: Not applicable.

Informed Consent Statement: Not applicable.

Data Availability Statement: The data presented in this study is available on request from the corresponding author.

Acknowledgments: We would like to thank for the support and ongoing collaboration with MultiMaterial-Welding AG and KVT-Fastening AG/Bossard during this study.

Conflicts of Interest: The authors declare no conflict of interest.

Appendix A

Table A1. FEM material definitions and contact properties.

Domain/Surface	Pin	Face Sheet	Honeycomb Cell	Foam
Element type	A 4-node thermally coupled tetrahedron, linear displacement and temperature	A 3-node thermally coupled triangular thin shell, finite membrane strain.	A 6-node thermally coupled triangular prism, linear displacement, and temperature.	
Density [tonne/mm^3]	1.34×10^{-9}	1.7×10^{-9}	1.3×10^{-9}	1.15×10^{-9}
Elastic modulus E [MPa]	Elastic modulus [MPa]/Temperature [K] 5500/303; 4051/323; 2357/348; 2085/373; 1784/398; 1587/423; 1314/473; 487/533	4550	400	300
Poisson's ratio ν [1]	0.35	0.35	0.33	0.2
Plastic properties	Stress σ [MPa]/Strain ε [1] 5/0; 10/0.005; 20/0.007; 30/0.01; 40/0.015; 50/0.017; 60/0.02; 70/0.025; 80/0.027; 90/0.03; 100/0.04; 110/0.045; 120/0.06; 130/0.07; 150/0.1	-	Stress σ[MPa]/Strain ε [1] 30/0 55/0.39	Stress σ[MPa]/Strain ε [1] 25/0 30/0.39

Table A1. *Cont.*

Domain/Surface	Pin	Face Sheet	Honeycomb Cell	Foam
Damage criteria	Ductile: displacement criteria	Hashin	Ductile: displacement criteria	Ductile: displacement criteria
Damage initiation	Equivalent plastic strain: Initiation strain [1]/ Temperature [K] 0.5/300 0.4/400 0.3/500	Tensile strength in the fibre direction X^T/Compressive strength in the fibre direction X^C/Tensile strength in the transverse direction Y^T/Compressive strength in the transverse direction Y^C/Longitudinal shear strength S^L/Transverse shear strength S^T 138/80/138/80/100/100	Initiation strain [1]: 0.5	Initiation strain [1]: 0.5
Damage evolution	Effective plastic displacement: 0.4	Energy: 1×10^{-9}	Effective plastic displacement: 0.001	Effective plastic displacement: 0.001
Thermal conductivity [W/mm/K]	Conductivity [W/mm/K]/Temperature [K] 0.36/300 0.5/400 0.6/500	0.1	0.1	0.1
Specific heat [J/K/tonne]	1.5×10^9	1×10^9	1×10^9	1×10^9
Inelastic heat fraction η_{pl} [1]	0.75	-	0.75	0.75
Friction coefficient μ_{fr} [1]	0.5			
Fraction of frictional work η_{fr} [1]	1			
Contact stiffness [MPa]	850			

References

1. Lambiase, F.; Balle, F.; Blaga, L.-A.; Liu, F.; Amancio-Filho, S.T. Friction-based processes for hybrid multi-material joining. *Compos. Struct.* **2021**, *266*, 113828. [CrossRef]
2. Staab, F.; Balle, F. Ultrasonic torsion welding of ageing-resistant Al/CFRP joints: Properties, microstructure and joint formation. *Ultrasonics* **2018**, *93*, 139–144. [CrossRef] [PubMed]
3. Thoppul, S.D.; Finegan, J.; Gibson, R.F. Mechanics of mechanically fastened joints in polymer–matrix composite structures—A review. *Compos. Sci. Technol.* **2009**, *69*, 301–329. [CrossRef]
4. Altmeyer, J.; dos Santos, J.; Amancio-Filho, S.T. Effect of the friction riveting process parameters on the joint formation and performance of Ti alloy/short-fibre reinforced polyether ether ketone joints. *Mater. Des.* **2014**, *60*, 164–176. [CrossRef]
5. Funari, M.F.; Greco, F.; Lonetti, P. Sandwich panels under interfacial debonding mechanisms. *Compos. Struct.* **2018**, *203*, 310–320. [CrossRef]
6. Funari, M.F.; Spadea, S.; Lonetti, P.; Lourenco, P.B. On the elastic and mixed-mode fracture properties of PVC foam. *Theor. Appl. Fract. Mech.* **2021**, *112*, 102924. [CrossRef]
7. Katnam, K.; Da Silva, L.; Young, T. Bonded repair of composite aircraft structures: A review of scientific challenges and opportunities. *Prog. Aerosp. Sci.* **2013**, *61*, 26–42. [CrossRef]
8. Ageorges, C.; Ye, L.; Hou, M. Advances in fusion bonding techniques for joining thermoplastic matrix composites: A review. *Compos. Part A Appl. Sci. Manuf.* **2001**, *32*, 839–857. [CrossRef]
9. Ueda, M.; Ui, N.; Ohtani, A. Lightweight and anti-corrosive fiber reinforced thermoplastic rivet. *Compos. Struct.* **2018**, *188*, 356–362. [CrossRef]
10. Fortier, V.; Brunel, J.-E.; Lebel, L.L. Fastening composite structures using braided thermoplastic composite rivets. *J. Compos. Mater.* **2019**, *54*, 801–812. [CrossRef]
11. Stavrov, D.; Bersee, H.E.N. Resistance welding of thermoplastic composites-an overview. *Compos. Part A Appl. Sci. Manuf.* **2005**, *36*, 39–54. [CrossRef]
12. Villegas, I.F.; van Moorleghem, R. Ultrasonic welding of carbon/epoxy and carbon/PEEK composites through a PEI thermoplastic coupling layer. *Compos. Part A Appl. Sci. Manuf.* **2018**, *109*, 75–83. [CrossRef]
13. Zweifel, L.; Brauner, C. Investigation of the interphase mechanisms and welding behaviour of fast-curing epoxy based com-posites with co-cured thermoplastic boundary layers. *Compos. Part A Appl. Sci. Manuf.* **2020**, *139*, 106–120. [CrossRef]
14. Ageorges, C.; Ye, L. Resistance welding of thermosetting composite/thermoplastic composite joints. *Adv. Mater.* **2006**, *32*, 1603–1612. [CrossRef]
15. Brauner, C.; Zweifel, L.; Dransfeld, C.; Rheme, M. Joining of multi-material assemblies by the MultiMaterial-Welding technology—Development of an online process monitoring system. In Proceedings of the International Conference and Exhibition on Thermoplastic Composites, Bremen, Germany, 30–31 October 2018.
16. Ahmed, T.J.; Stavrov, D.; Bersee, H.E.N.; Beukers, A. Induction welding of thermoplastic composites—An overview. *Compos. Part A Appl. Sci. Manuf.* **2006**, *37*, 1638–1651. [CrossRef]

17. Villegas, I.F.; Palardy, G. Ultrasonic welding of CF/PPS composites with integrated triangular energy directors: Melting, flow and weld strength development. *Compos. Interfaces* **2016**, *24*, 515–528. [CrossRef]
18. Levy, A.; Le Corre, S.; Villegas, I.F. Modeling of the heating phenomena in ultrasonic welding of thermoplastic composites with flat energy directors. *J. Mater. Process. Technol.* **2014**, *214*, 1361–1371. [CrossRef]
19. Bhudolia, S.K.; Gohel, G.; Leong, K.F.; Islam, A. Advances in Ultrasonic Welding of Thermoplastic Composites: A Review. *Materials* **2020**, *13*, 1284. [CrossRef]
20. Jongbloed, B.; Teuwen, J.; Palardy, G.; Villegas, I.F.; Benedictus, R. Continuous ultrasonic welding of thermoplastic composites: Enhancing the weld uniformity by changing the energy director. *J. Compos. Mater.* **2019**, *54*, 2023–2035. [CrossRef]
21. Brauner, C.; Nakouzi, S.; Zweifel, L.; Tresch, J. Co-curing behaviour of thermoset composites with a thermoplastic boundary layer for welding purposes. *Adv. Compos. Lett.* **2020**, *29*, 2633366X20902777. [CrossRef]
22. Larsen, L.; Görick, D.; Engelschall, M.; Fischer, F.; Kupke, M. Process data driven advancement of robot-based continuous ultrasonic welding for the dust-free assembly of future fuselage structures. In Proceedings of the International Conference and Exhibition on Thermoplastic Composites, Bremen, Germany, 13–15 October 2020.
23. Tolunay, M.N.; Dawson, P.R.; Wang, K.K. Heating and bonding mechanisms in ultrasonic welding of thermoplastics. *Polym. Eng. Sci.* **1983**, *23*, 726–733. [CrossRef]
24. Zhang, C.; Li, L. A Coupled Thermal-Mechanical Analysis of Ultrasonic Bonding Mechanism. *Met. Mater. Trans. A* **2009**, *40*, 196–207. [CrossRef]
25. Levy, A.; Le Corre, S.; Poitou, A. Ultrasonic welding of thermoplastic composites: A numerical analysis at the mesoscopic scale relating processing parameters, flow of polymer and quality of adhesion. *Int. J. Mater. Form.* **2012**, *7*, 39–51. [CrossRef]
26. Lee, W.I.; Springer, G.S. A Model of the Manufacturing Process of Thermoplastic Matrix Composites. *J. Compos. Mater.* **1987**, *21*, 1017–1055. [CrossRef]
27. MATLAB—Tracking and Motion Estimation. Available online: https://ch.mathworks.com/help/vision/tracking-and-motion-estimation.html (accessed on 7 July 2021).
28. ABAQUS Analysis User's Manual, Version 6.6. 2006. Available online: https://www.worldcat.org/title/abaqus-analysis-users-manual-version-66/oclc/1076085564?referer=di&ht=edition (accessed on 27 August 2021).
29. Sayles, R.S.; Thomas, T.R. Surface topography as a nonstationary random process. *Nature* **1978**, *271*, 431–434. [CrossRef]
30. Yang, F.; Pitchumani, R. Interlaminar contact development during thermoplastic fusion bonding. *Polym. Eng. Sci.* **2002**, *42*, 424–438. [CrossRef]
31. Loos, A.C.; Dara, P.H. Processing of Thermoplastic Matrix Composites. In *Review of Progress in Quantitative Nondestructive Evaluation*; Thompson, D.O., Chimenti, D.E., Eds.; Springer: Boston, MA, USA, 1987; pp. 1257–1265.
32. Mantell, S.C.; Springer, G.S. Manufacturing Process Models for Thermoplastic Composites. *J. Compos. Mater.* **1992**, *26*, 2348–2377. [CrossRef]

materials MDPI

Article

The Mechanical Characterization of Welded Hybrid Joints Based on a Fast-Curing Epoxy Composite with an Integrated Phenoxy Coupling Layer

Lucian Zweifel [1,*], Klaus Ritter [2] and Christian Brauner [1]

1 Institute of Polymer Engineering, University of Applied Sciences and Arts Northwestern Switzerland (FHNW), Klosterzelgstrasse 2, 5210 Windisch, Switzerland; christian.brauner@fhnw.ch
2 Huntsman Advanced Materials, Klybeckstrasse 200, 4057 Basel, Switzerland; klaus_ritter@huntsman.com
* Correspondence: lucian.zweifel@fhnw.ch

Abstract: The joining of composites mostly relies on traditional joining technologies, such as film or paste adhesives, or mechanical fasteners. This study focuses on the appealing approach of using standard thermoplastic welding processes to join thermosets. To achieve this, a thermoplastic coupling layer is created by curing with a thermoset composite part. This leads to a functional surface that can be utilized with thermoplastic welding methods. The thermoplastic coupling layer is integrated as a thin film, compatible with the thermoset resin in the sense that it can partially diffuse in a controlled way into the thermoset resin during the curing cycle. Recent studies showed the high affinity for the interphase formation of poly hydroxy ether (phenoxy) film as coupling layer, in combination with a fast-curing epoxy system that cures within 1 min at 140 °C. In this study, an investigation based on resistance and ultrasonic welding techniques with different testing conditions of single-lap shear samples (at room temperature, 60 °C, and 80 °C) was performed. The results showed strong mechanical strengths of 28.9 MPa (±0.7%) for resistance welding and 24.5 MPa (±0.1%) for ultrasonic welding, with only a minor reduction in mechanical properties up to the glass transition temperature of phenoxy (90 °C). The combination of a fast-curing composite material with an ultra-fast ultrasonic joining technology clearly demonstrates the high potential of this joining technique for industrial applications, such as automotive, sporting goods, or wind energy. The innovation allowing structural joining performance presents key advantages versus traditional methods: the thermoplastic film positioning in the mold can be automated and localized, joint formation requires only a fraction of a second, and the joining operation does not require surface preparation/cleaning or structure deterioration (drilling).

Keywords: phenoxy; welding technology; interphase; ultrasonic welding; weld properties; advanced composites

Citation: Zweifel, L.; Ritter, K.; Brauner, C. The Mechanical Characterization of Welded Hybrid Joints Based on a Fast-Curing Epoxy Composite with an Integrated Phenoxy Coupling Layer. *Materials* **2022**, *15*, 1264. https://doi.org/10.3390/ma15031264

Academic Editor: Patricia Krawczak

Received: 19 January 2022
Accepted: 4 February 2022
Published: 8 February 2022

Copyright: © 2022 by the authors. Licensee MDPI, Basel, Switzerland. This article is an open access article distributed under the terms and conditions of the Creative Commons Attribution (CC BY) license (https://creativecommons.org/licenses/by/4.0/).

1. Introduction

Reliable and cost-effective joining technologies for fiber-reinforced composite materials provide a great potential to significantly reduce weight, fuel consumption, and, consequently, CO_2 emissions [1–5]. Therefore, it is essential to develop and implement new joining technologies to further improve the manufacture and assembly of structural composite parts. The use of thermoplastic joining processes has certain benefits in joining thermoset composite structures. So far, joining similar and dissimilar composite has mostly relied on traditional joining technologies such as adhesive bonding or mechanical fasteners, which both have distinct drawbacks. Joining with adhesives results in high costs and long process times, whereas mechanical fasteners introduce weakness into the structure (stress concentrations) due to the necessity of drilling holes, where the continuous fiber reinforcement is disrupted. In contrast, thermoplastic welding offers the unique ability of melting and reprocessing compared to thermosets, which cannot be re-melted after cross-linking. Additionally, thermoplastic welding allows for fast processing speeds without significant surface preparation efforts, resulting in strong and dependable mechanical performance [4,6].

Within the last three decades, thermoplastic welding processes have become increasingly interesting in research, as well as for use in industry moving towards primary structures in aerospace [1]. Here, resistance welding [6–12], induction welding [1,2], ultrasonic welding [6,13–16], and laser welding [6] have proven their potential as reliable joining technologies. Since 2001, the glass fiber-reinforced polyphenylene sulfide (PPS) J-Noses for the Airbus A340-500/600 and A380-800 are resistance-welded in series production, resulting in a weight reduction of greater than 20% [17,18]. In 2010, Fokker Aerostructures designed and developed the Gulfstream G650 tail section and industrialized the induction welding method for the rudder and the elevator using carbon fiber-reinforced PPS, reducing the weight of the components by 25% compared to traditional materials [12,19]. Within the EU's CleanSky 2 program, ultrasonic welding was utilized to spot-weld short fiber-reinforced thermoplastic clips to the skin and stringers of a multifunctional fuselage demonstrator [20,21]. Furthermore, MM-Welding® has presented innovative fastening techniques to create cost-effective joining of fixation elements in differently structured materials (porous materials, sheet structures, sandwich materials, injection-molded materials, and pressure moldings) via ultrasonic energy in the automotive field [22,23]. Recently, robotic-based, continuous welding of thermoplastic structures was introduced at DLR Augsburg [24], showing the potential towards the energy-efficient automation of welding processes. Each technology has shown distinct limitations, such as resistance welding often using a metal wire mesh, which is ideal for insulating materials, such as glass fibers, but leads to energy leakage with carbon-reinforced composites due to conductivity [12]. The above-mentioned applications focused on thermoplastic composites for high-performance applications using materials such as fiber-reinforced PPS, polyetheretherketone (PEEK), polyetherketoneketone (PEKK), polyaryletherketone (PAEK), and polyetherimide (PEI). However, an interesting question is whether it is possible to also apply welding to dissimilar composites, e.g., thermoplastic and thermoset composite combinations, as a composite-friendly alternative to current assembling procedures. Here, the attractive concept of using the thermoplastic welding process for thermosets was introduced [25] by including a thermoplastic-rich layer during the curing process of a thermoset composite. The so-called thermoplastic coupling layer acts as a functionalized, or 'weldable', surface (see Figure 1).

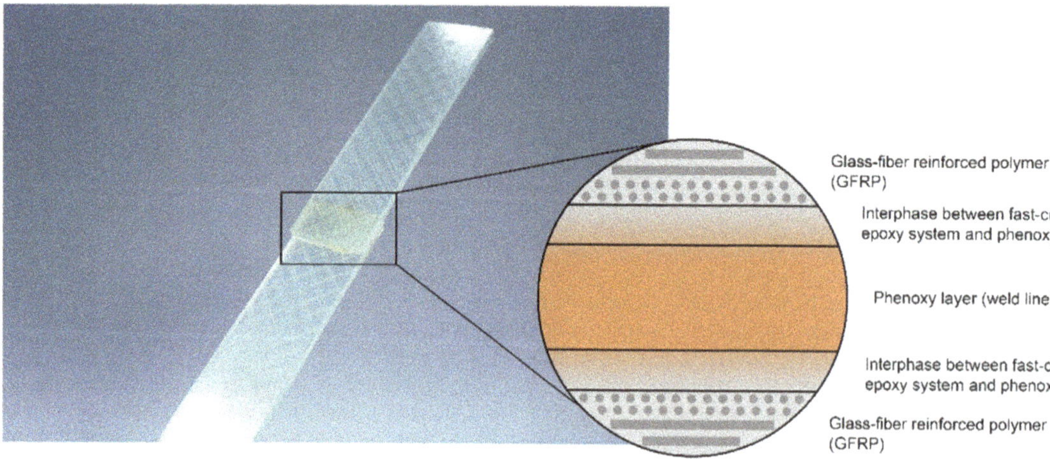

Figure 1. Concept of weldable hybrid joints based on a fast-curing epoxy composites with an integrated poly hydroxy ether (phenoxy) coupling layer.

The affinity between thermoset and thermoplastic is the crucial element for creating joints with high strength [13,26–28]. A gradient interphase forms between the reactive epoxy resin and the thermoplastic material, whereby the two components partially dissolve, diffuse, and, finally, decompose due to a reaction-induced phase separation [29–31]. The overall composition or cross-link density of the material varies from location to location at the macroscopic level [32]. The decomposed morphology enables strong mechanical interlocking for subsequent load transfer [28]. Recent studies have proposed PEI as a suitable candidate for interphase formation in aerospace-related applications [9,13,26,28,30,33–35]. In a previous study, an identical concept was applied to automotive-related applications with a fast-curing epoxy system (Araldite LY3585/Aradur 3475) [27]. It was possible to establish an interphase between poly hydroxy ether (phenoxy) and the fast-curing epoxy system within a fraction of a minute, leading to a high mechanical performance of single-lap shear samples joined by resistance welding, with an average strength of 25.3 MPa [27]. For automotive applications, comparable structural adhesives are listed in Table 1. The comparison between structural adhesive bonding and thermoplastic welding strengthens the potential of the latter. Therefore, the joining of thermosets via a phenoxy boundary layer is a reliable, cost-efficient technique, which offers the possibility of creating lightweight design concepts.

Table 1. Comparison of commercial structural adhesive solutions utilized in the automotive industry (GFRP = glass fiber-reinforced polymer; CFRP = carbon fiber-reinforced polymer, LSS = lap shear strength).

Structural Adhesive	Curing Condition	Adherend	Testing Temperature	LSS (MPa)
SikaPower® 1200 [36]	4 h at 70 °C	GFRP	23 °C	20
3M™ Structural Adhesive SA9820 [37]	24 h at RT followed by 30 min at 170 °C	Aluminum	23 °C	20
3M™ Structural Adhesive SA9820 [37]	24 h at RT followed by 30 min at 170 °C	Aluminum	80 °C	13
Araldite® 2012 [38]	16 h at 40 °C	CFRP	23 °C	14.5
SikaPower®-1277 [39]	2 weeks at 23 °C	Steel	23 °C	28

Within this study, an investigation using a fast-curing epoxy resin and phenoxy was conducted with resistance and ultrasonic welding techniques whereby the lap shear strength (LSS) was evaluated using single-lap shear tests with different testing conditions (at room temperature, 60 °C, and 80 °C). In addition, the failure mechanism of fractured samples combined with optical microscopy of the microstructure was analyzed. The objective was the adaptation from lab-scale trials to industrial-scale manufacturing methods utilizing dynamic fluid compression molding (DFCM), which is an efficient compression molding process whereby a one-shot phenoxy film integration was tested.

2. Materials and Methods

2.1. Materials

Phenoxy polymers are commercially available in different forms such as emulsion, powders, films, and granulates. In this study, the phenoxy grade PKHH (Huntsman Advanced Materials, Basel, Switzerland), a grade for extrusion, was selected as the most promising candidate for welding processing. The structure of this polymer (Figure 2) shows amorphous thermoplastic properties, such as rigidity, thermal and chemical stability, and adhesion strength. The rigidity and thermal stability result from the aromatic compounds (marked in blue), the chemical stability of the oxygen atoms in the main chain (marked in yellow), and the adhesion strength of the hydroxyl groups (marked in green). The PKHH material has a molecular weight of 52,000 g/mol [40], and compared to available phenoxy resins, such as PKHB (32,000 g/mol) and PKFE (60,000 g/mol), it is positioned in the upper range [40]. Two different film thicknesses (75 μm and 125 μm) were used for parametric studies and were supplied by Huntsman Advanced Materials, Basel, Switzerland.

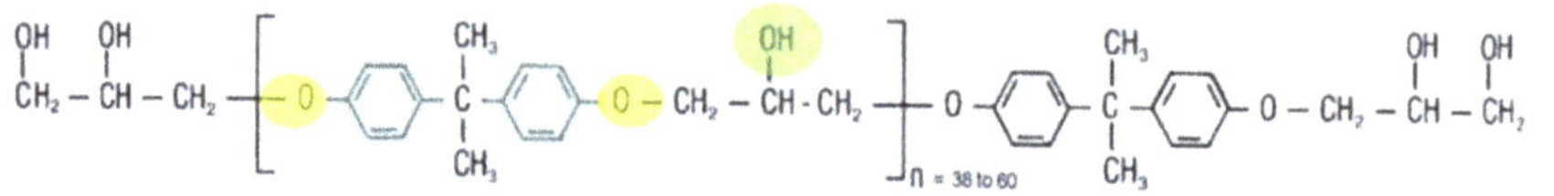

Figure 2. The polymer structure of poly hydroxy ether (phenoxy).

In this study, an epoxy resin system (Araldite LY3585/Aradur 4375) provided by Huntsman Advanced Materials (Basel, Switzerland) was used. The system is widely applied in the mass production of structural automotive components and is notable for its short curing cycles and final glass transition temperature of approximately 120 °C. The mentioned system is typically used for high-pressure resin transfer molding (HP-RTM), wet compression molding, and dynamic fluid compression molding (DFCM) applications, with a curing cycle of 2 min at 115 °C and 1 min at 140 °C, respectively. The DFCM process combines the speed of wet compression molding with the quality of HP-RTM without sacrificing geometric complexity [41,42]. With the usage of vacuum and dynamic mold pressure, void-free impregnations are achieved [41]. The epoxy resin system is based on bisphenol-A-diglycidylether as resin and 1,3-cyclohexanedimethanamine and methyldiethanolamine as hardeners. A kinetic model has been developed by the authors using the modified Kamal-Sourour model to describe the curing process [27], which will also be used in this study.

2.2. Manufacturing Methods

Composite plates were manufactured using the DFCM process. The plates were cured for 1 min at 140 °C, resulting in a part production time of 1.5 min. The process consisted of the following steps: First, the mold was heated to 140 °C. Second, the defined amount of resin mix was spread onto the preform, outside of the mold. Third, the preform (with the resin mix on top) was placed into the mold. Fourth, the mold was partially closed, and the vacuum was drawn. After 10 s, the mold was fully closed, and after an additional 60 s of curing with an applied pressure of 30 bars, the mold was opened, and the part was removed. The laminate consisted of two main types of fabrics: two layers of a canvas weave, glass fiber fabric with an aerial weight of 200 g/m^2 (0/90, Porcher Industries, Erbach, Germany) as outer layers, and six layers of a glass fabric with a higher aerial weight of 600 g/m^2 (EBX 600, 45/−45, Saati Composites, Appiano Gentile, Milano, Italy) as inner layers. Thus, two different layups were used, with the labeling L1 ($0/-90_{GF200}$, $45/-45_{GF600}$, $-45/45_{GF600}$, and $45/-45_{GF600})_s$ and L2 ($45/-45_{GF600}$, $-45/45_{GF600}$, and $45/-45_{GF600})_s$. Glass fiber fabrics were selected because they allow a clear visual inspection of the sample after manufacturing, joining, and testing. Different types of fabrics were chosen, as the weave pattern significantly influences crack propagation and LSS [7,43]. The mold dimension was 530 × 530 mm^2, which theoretically results in approximately 95 single-lap shear coupons.

The sensitive element of the manufacturing was the introduction of the thermoplastic film into the process in a reliable, robust way that could be later adapted by industrialization. Thus, the thermoplastic coupling layer was added on the lower side of the tool with a thickness of 125 μm, which provides material for interphase formation (approximately 13 μm at a 100 °C curing temperature [27]) and welding. It was assumed that the considerably small thickness of the phenoxy film does not influence the residual stress of the laminate stacking sequence. The phenoxy film was partially integrated in joining areas (Figure 3) at positions where an actual welding was later conducted (applying the right material in the right place). In this study, the phenoxy film was directly taped into the mold before the mold was heated up.

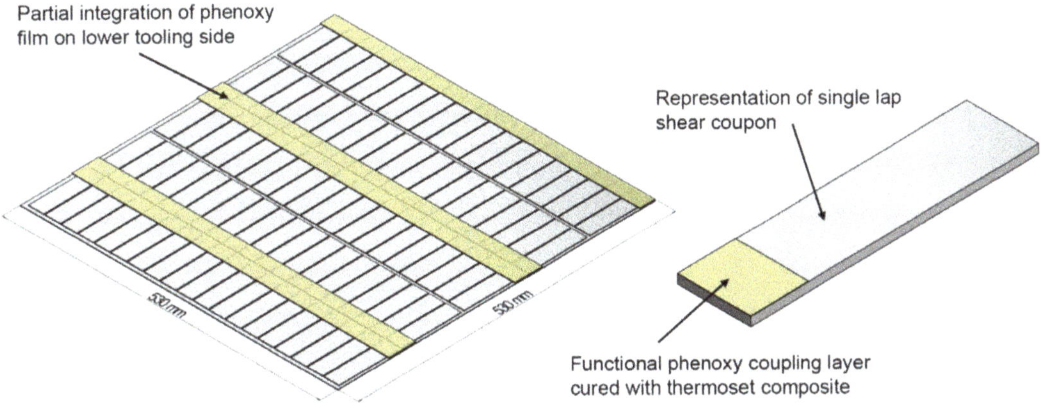

Figure 3. Schematic lap shear coupon array with partial integration of the poly hydroxy ether (phenoxy) film on the lower tooling side.

2.3. Welding and Test Methods

Two welding methods, namely, resistance welding and ultrasonic welding, were used and compared in this study. The welded joints were mechanically tested following the ISO 4587 standard using a Zwick 100 kN universal tensile test machine (Zwick Roell, Ulm, Germany) with a testing speed of 2 mm/min to determine the apparent LSS. The LSS was calculated as the maximum load divided by the overlap area. No surface preparation of the overlapping area was applied. As phenoxy grades are sensitive to water absorption, the samples were dried for 5 h at 80 °C [42]. Furthermore, fractured specimens were analyzed by visual examination of the fracture area according to DIN EN ISO 10365. First, a comparative study was performed with resistance and ultrasonic welding, whereby the process parameters were iteratively optimized to achieve maximum mechanical properties. Here, both welding methods have been assessed by determining of the LSS at room temperature, followed by an analysis of the fractured area. Second, the ultrasonic-welded samples were tested at higher temperatures (at 60 °C and 80 °C). The testing temperatures were chosen due to their relevance in the automotive fields, as well as to test the limit of mechanical performance below the glass transition temperature of phenoxy (~91 °C). For each parametric configuration, five samples were welded and tested.

2.3.1. Resistance Welding

The custom-built resistance welding setup, developed in a previous study [8], was used to join single-lap shear test specimens. The welding process parameters were experimentally optimized in a preliminary study resulting in a power of 25 kW/m^2, a pressure of 1 MPa, and a total welding time of 80 s, followed by 180 s of cooling time while the pressure was maintained. Detailed investigations of the process parameters coupled with in-situ process monitoring were described in [8,27]. Further, a stainless steel heating element with a wire diameter of 36 μm, a mesh size of 50 μm, and an open area of 33.8% was used [27]. The heating elements were cut to dimensions of 12.5 mm wide and 50 mm long.

2.3.2. Ultrasonic Welding

In this study, a pneumatic-driven ultrasonic welder (Branson 2000Xc, Emerson Automation Solutions, Baar, Switzerland) was used. The equipment monitored force, displacement, amplitude, frequency, and energy, with a sampling rate of 100 Hz. A rectangular sonotrode made of titanium alloy (Ti-6Al-4V) with a welding area of 19 × 38 mm^2 was used. The sonotrode has a gain factor of 1.5, resulting in a maximum amplitude of 76 μm at 100% power output. The optimal process settings for ultrasonic welding were experimentally determined in a preliminary study, resulting in the following parameters: trigger force,

50 N; ultrasonic active, 0.5 s; amplitude, 100%; weld pressure, 250 kPa; hold pressure, 350 kPa; hold time, 2 s; and operating frequency, 20 kHz. The hold process after welding was important to ensure an ideal consolidation quality by cooling the joining area.

One method to generate controlled heat in the welding zone is to introduce energy directors in the ultrasonic welding process to concentrate the heat generation within the boundary of the two constituents through a combination of surface friction and viscoelastic heating [44]. For all the welded joints, a 75 μm thick, flat phenoxy energy director was used to concentrate heat generation at the welding interface. Flat energy directors are neat resin films with a slightly larger size of 30 × 30 mm^2 compared to the joining area. The films were placed between the adherends before welding. Flat energy directors led to similar results compared to more traditionally molded energy directors [13,45–47]. Recent studies analyzed energy director-less joints compared to flat energy directors, whereby a significant reduction of LSS was apparent [33]. Furthermore, the influence of the final weld line thickness between the two adherends on the resulting LSS is significant, and it is comparable to traditional adhesive technology. It has been shown experimentally by many authors that the joint strength decreases as the weld line thickness increases due to the introduction of bending stresses [7,48,49]. Consequently, there is an optimum value of weld line thickness for which the LSS is maximized. Therefore, a second energy director thickness of 125 μm was used for one configuration of ultrasonic welding.

3. Results and Discussion

3.1. Analysis of Phenoxy Film Integration during Manufacturing

Figure 4 presents the concept of the one-shot phenoxy film integration before and after the preheating phase of the DFCM process. The fixation of the film was achieved by using a temperature-resistant tape that secured any movement in plane direction. During the heat up of the mold to the temperature of 140 °C for the DFCM process, the film was stable in position, without shrinkage effects (see Figure 4b). Furthermore, no overflow of resin on the welding surface was visible.

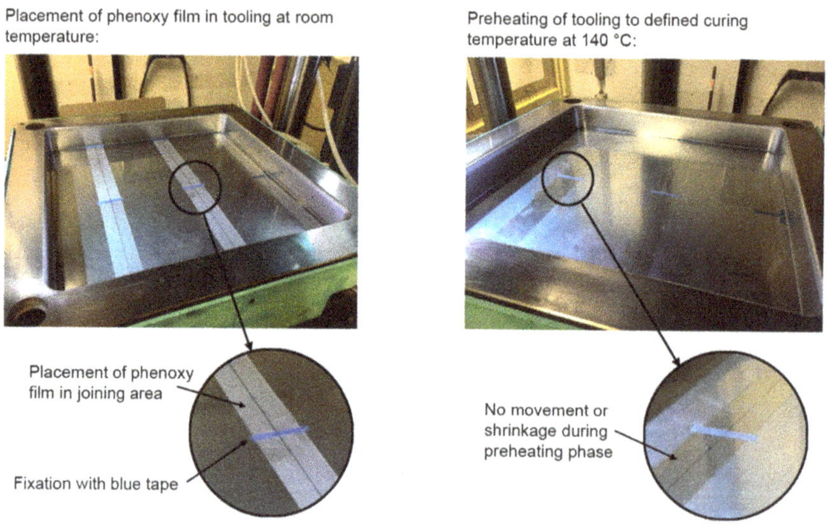

(a) (b)

Figure 4. One-shot phenoxy film integration during the preheating phase of the dynamic fluid compression molding (DFCM) process. (**a**) Placement of the phenoxy film in the tooling at room temperature. (**b**) Preheating of the tooling to a defined curing temperature of 140 °C.

After curing the composite plate, the phenoxy film was fully integrated in the thermoset glass fiber-reinforced polymer (GFRP) part, showing no indication of separation and no change in position or distortion from the vacuum, and therefore demonstrating the high affinity of the two components [27]. Despite the high curing temperature, and thus the fast polymerization of the thermoset within less than 1 min, the phenoxy film partially dissolved and formed an interphase with the epoxy resin, as shown in [27] with Raman spectroscopy. The actual time for interphase formation is much less than 60 s, as the highest diffusion rate prevails in the beginning of the curing. As the chain mobility reduces with a higher degree of curing, the diffusion rate drops rapidly, eventually leading to the maximum interphase thickness within the first 10 s of curing [28,30,31,50]. Voleppe et al. [51] showed that the penetration at the front of the thermoset continued beyond phase separation, whereby both mechanisms overlapped. The measurement of the phenoxy coupling layer was 85 μm after curing (see Figure 5), whereas the nominal film thickness prior to curing was 125 μm. The actual thickness after curing cannot be measured by optical microscopy as the contrast between the two components is not sufficient to provide clear distinction due primarily to their similar chemical structure, as both phenoxy and the epoxy systems are based on bisphenol A, which results in a similar optical appearance and spectroscopy response [27]. The layup resulted in an average laminate thickness of 3.05 mm with a standard deviation of ±3%. The quality was consistent for all four manufactured plates, with the absence of porosity and a constant fiber volume content. The cross-sections shown in Figure 5 present the similar appearances of the interphase shown in [27]. Furthermore, no porosities were found within the phenoxy/GFRP interphase.

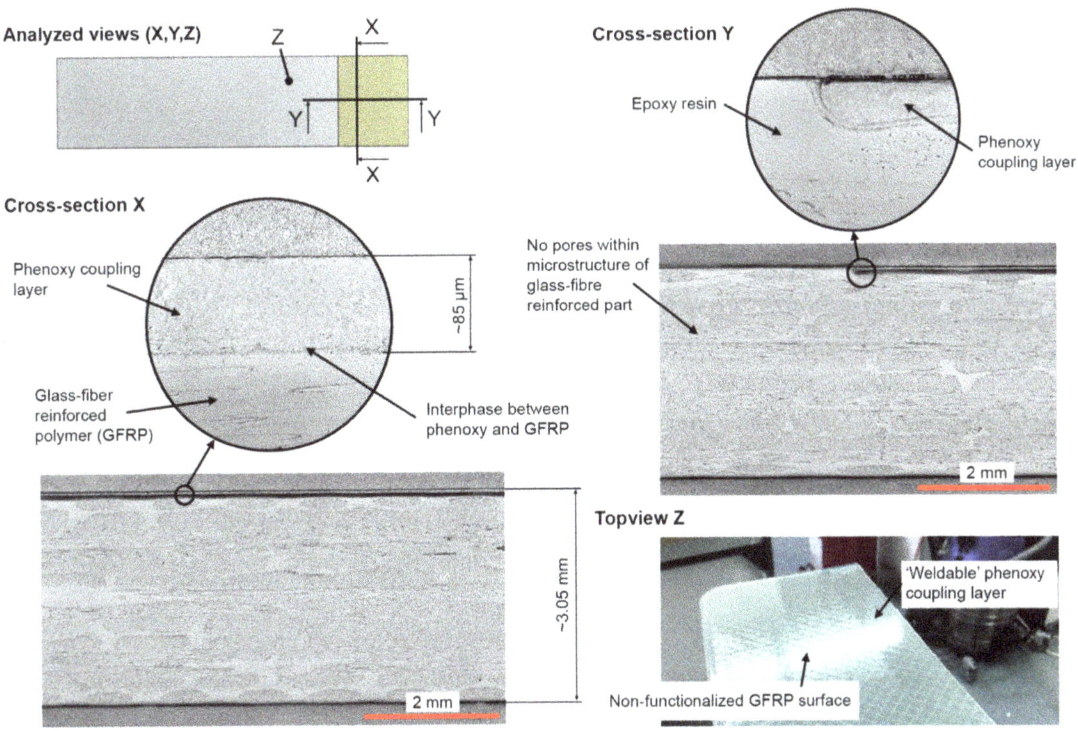

Figure 5. Cross-sectional optical micrographs and visual appearance of the phenoxy/GFRP plate samples: X: Optical representation of the phenoxy coupling layer showing a high affinity to the epoxy resin; Y: Resulting transition between the cured epoxy resin and the integrated phenoxy coupling layer; Z: Visual appearance of the 'weldable' and 'non-weldable' surface on a phenoxy/GFRP sample.

3.2. Welding and Evaluation of Mechanical Performance

In the following section, the welding results are presented. First, exemplary online process monitoring results for the ultrasonic (Figure 6a) and resistance welding (Figure 6b) processes are shown to provide insight into the welding process. In a second step, the visual appearance is compared. Third, the results of the mechanical tests are shown by analyzing the fracture surface, shear stress-displacement curves, and the resulting LSS. For the resistance welding, a welding pressure of 1 MPa was selected. During the process, the phenoxy was heated by the Joule effect [9–11]. After reaching the glass transition temperature of approximately 90 °C, the phenoxy interlayer starts to soften. Reaching the glass transition point is visible in a drop of applied pressure after 20–30 s. The total welding time was 80 s, with a further consolidation during cooling for 180 s. Therefore, the total process time was around 260 s. In comparison to the resistance welding process, the ultrasonic welding process was faster with a total time of 3 s. In this case, the welding process takes 0.5 s using an amplitude of 76 μm, followed by a consolidation time of 2 s.

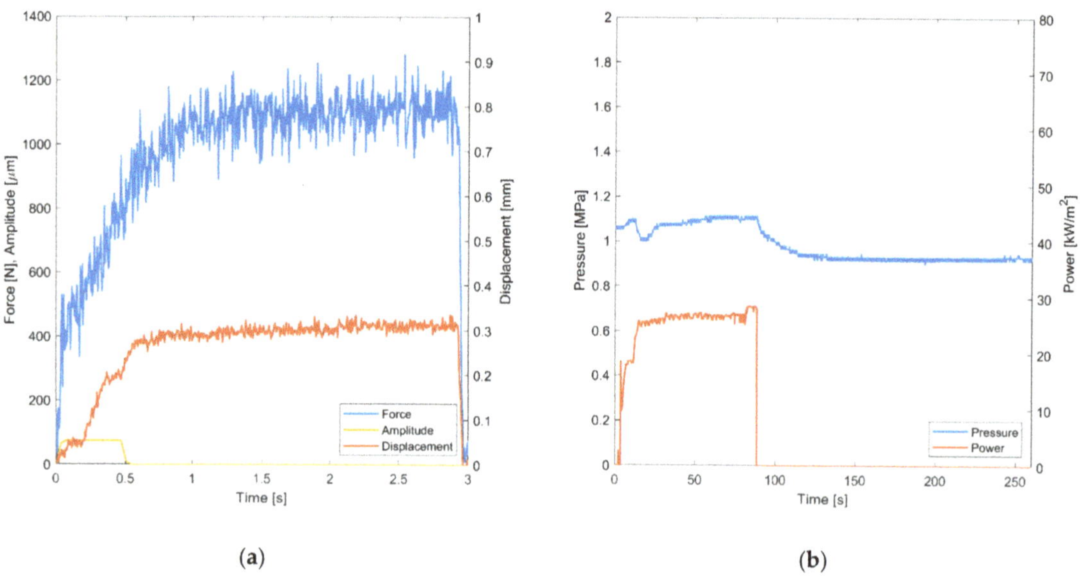

Figure 6. Comparison of online process monitoring for both thermoplastic welding processes. (**a**) Force, amplitude and displacement of the designated ultrasonic welding process sampled at 100 Hz. (**b**) Pressure and power of the designated resistance welding process sampled at 100 Hz.

Both welding technologies were optimized successfully with proper visual appearance of the samples (see Figure 7a). Only a slight flow of the thermoplastic material was apparent for both resistance and ultrasonic welding. For ultrasonic welding with the 75 μm thick energy directors, the final weld line thickness was approximately 118 μm with a standard deviation of 7 μm (±6.5%). Thereby, the weld line quality was satisfactory and consistent, with only minor porosities for samples joined by ultrasonic welding, and it was comparable to other research [13,52]. The micrographs underline the ideal process parameters and settings to exploit the full potential of the joining technology. Single-lap shear joints are prone to peel stresses and excessive shear stress on the tip of the weld line [49,53]. Consequently, excessive material that was squeezed out of the joining area during welding was removed before testing, primarily due to maintaining a consistent joint design throughout the joined coupons because the fillets formed by the squeeze flow did not show the exact characteristic for each weld performed.

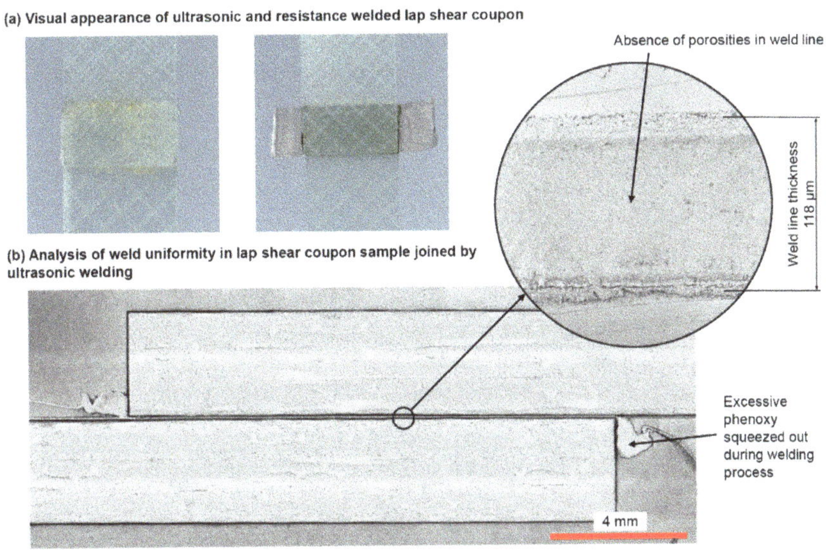

Figure 7. (**a**) Visual appearance of the ultrasonic- and resistance-welded lap shear coupons. (**b**) Micro-polished cross-section of lap shear coupon joined by ultrasonic welding.

The samples were tested to derive the LSS. In Figure 8, the tested samples are presented in detail to highlight the different failure types. As visible, the failure types were consistent for all samples where the failure occurred mostly within the GFRP (adherend) and not as an adhesive or cohesive failure. This confirms that the process parameters were selected properly for both welding methods. In the case of too low temperatures, this would lead to a cohesive failure between the samples. In the case of too high temperatures, this would lead to a color change (thermal degradation) of the welded zone and to an adhesive failure. In the thermogravimetric analysis measurement published in [54], the thermal stability of phenoxy was characterized by showing an onset of decomposition at 340 °C. It was observed from the results that the dominant composite failure is the ideal case because the joining area is stronger than the composite itself. Figure 7 supports this argument, as no porosities or other defects were present in the optical micrograph of the sample cross-section. Furthermore, there were only minor unwelded areas detected within the overlap, which eventually explains the drop in LSS of some samples at higher testing temperatures.

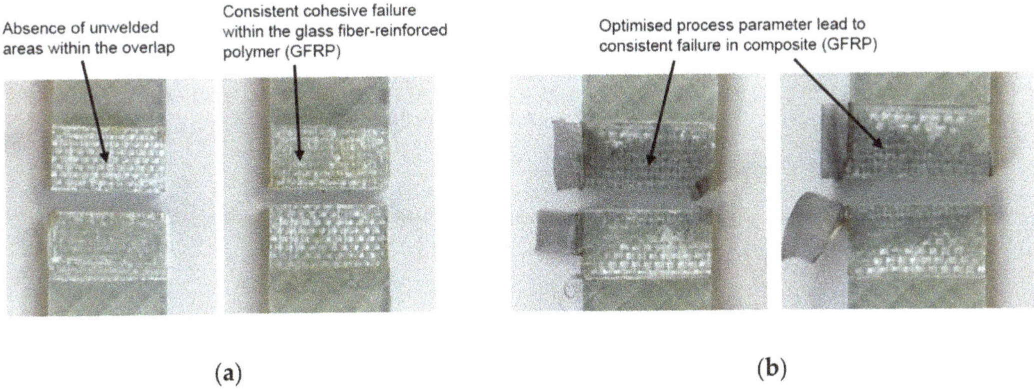

Figure 8. Exemplary fractured samples for (**a**) ultrasonic welding and (**b**) resistance welding.

In Figure 9, the shear stress-displacement curves are shown for the two welding methods. It is visible that both welding methods show a linear, consistent behavior until 15 MPa. After this, failure of the specimens starts. The typical failure is not brittle, and it seems that the damage tolerance is higher compared to the typical failure of an adhesive-bonded specimen [48,49].

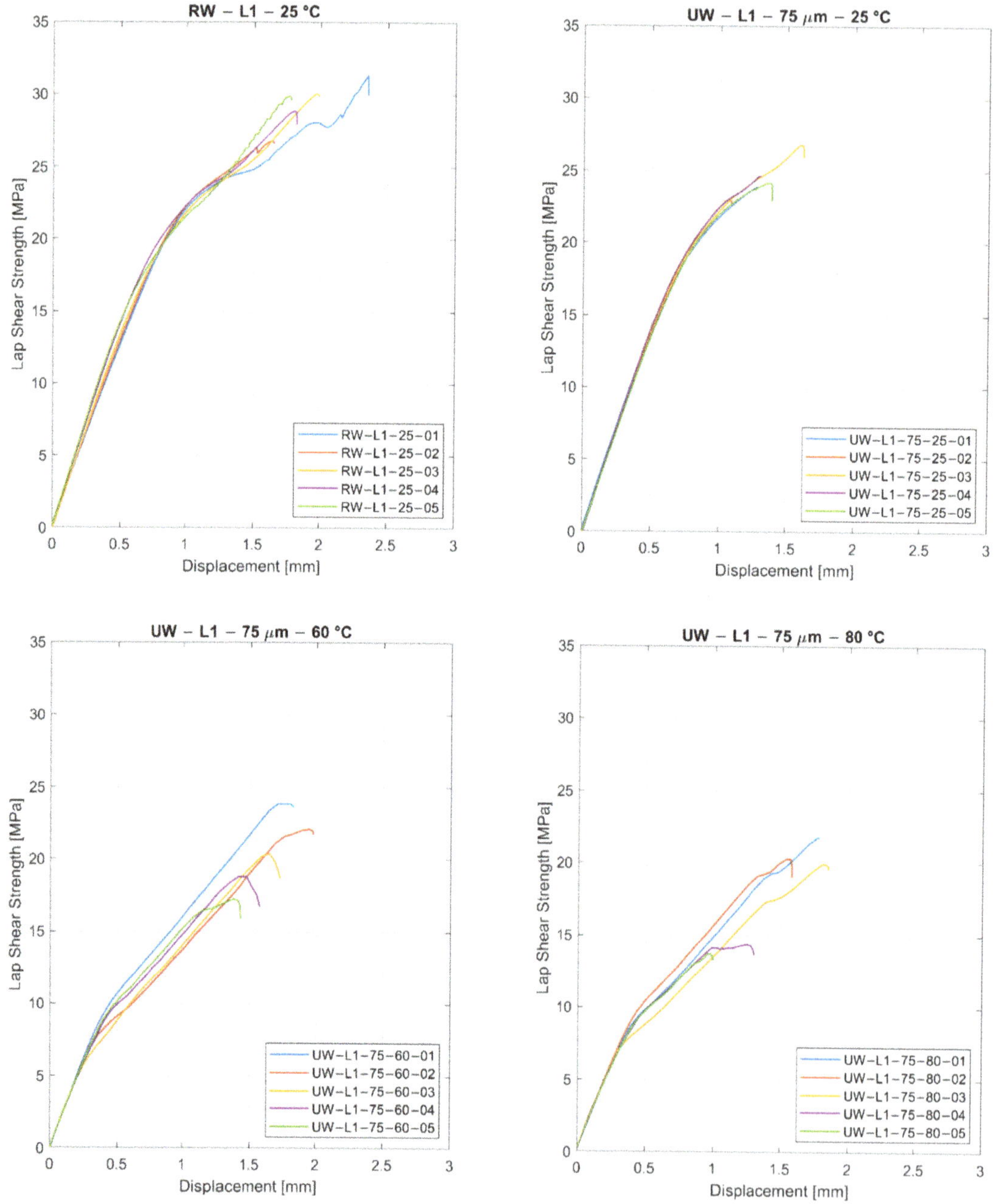

Figure 9. Comparison of stress versus displacement depending on the welding method (RW = resistance welding and UW = ultrasonic welding), layup, energy director, and testing temperature.

Table 2 illustrates the average LSS between two welded phenoxy/GFRP coupons with their respective thermoplastic boundary layers for resistance welding (RW) and ultrasonic welding (UW). Using resistance welding, the maximum shear strength was 28.5 MPa with a standard deviation of 2.3 MPa. The comparable ultrasonic welded specimens (L1, 75 µm, 25 °C) failed at 23.8 MPa with a standard deviation of 0.6 MPa. In comparison, the resistance-welded samples have a 19% higher strength, which was explained by the incorporation of the steel mesh that acts as a reinforcement. The mesh improves the shear strength because it influences the crack propagation, whereby micro-cracks are deflected [7]. As is visible in the results, the thickness of the phenoxy layer (energy director) has only a minor influence on the shear strength, especially because the final weld line thickness is more influenced by the pressure applied during the weld (with dependency on squeeze flow). Both 75 µm and 125 µm energy directors led to a similar weld line thickness between 110–120 µm. In [27], it was shown by hot stage microscopy and Raman microscopy that for isothermal curing conditions at 100 °C with similar materials results in an interphase with a thickness of 13 µm with a linear concentration gradient. Furthermore, the change in layup (L1, L2) shows a significant drop in apparent LSS as the propagation of micro-cracks depends on the weave pattern located on the outer layer of the composite.

Table 2. Resulting lap shear strength (LSS) depending on welding technology (RW = resistance welding, UW = ultrasonic welding), layup, energy director, and testing temperature.

Welding	Layup	Energy Director	Testing Temperature	LSS (MPa)	Standard Deviation (MPa)
RW	L1	-	23 °C	28.5	2.3
UW	L1	75 µm	23 °C	23.8	0.6
UW	L1	125 µm	23 °C	24.4	0.4
UW	L2	75 µm	23 °C	20.8	2.2
UW	L1	75 µm	60 °C	20.1	2.4
UW	L1	75 µm	80 °C	18.3	3.3

In a second step, ultrasonic-welded samples were tested at 60 °C and 80 °C. The glass transition temperature of the utilized phenoxy grade PKHH is 91 °C (further data about the temperature-dependent stiffness was published in [54]). In [54], three-point bending tests using a DMA Q800 (Dynamic Mechanical Analysis) were performed on a phenoxy PKHB sample. The results have shown a very linear decrease of the storage modulus until the glass transition onset of 91 °C was reached [54]. The resulting LSS was influenced by the temperature but remained at a considerable high strength, with minor reductions for 60 °C (3.4%) and 80 °C (12%). Here, small defects within the welding interface, such as unwelded areas or weld line thickness inconsistency, became more severe and eventually resulted in a higher scattering of the apparent maximum LSS. In comparison to the tests at room temperature, the displacement at maximum shear strength slightly increased at higher testing temperatures due to the more ductile behavior of the welding area. Furthermore, it was possible to allocate small, unwelded areas within the samples 'UW-L1-75-80-04' and 'UW-L1-75-80-05', which provides a reasonable explanation for the drop in maximum LSS compared to the other samples tested at 80 °C.

4. Conclusions

In this paper, the experimental assessment of the manufacturing and mechanical characterization of phenoxy/GFRP hybrid joints based on ultrasonic and resistance welding was presented. Besides the successful phenoxy film integration during manufacturing, different welding configurations were analyzed: influence of welding method on mechanical performance, effect of weave pattern on lap shear strength (LSS), and reduction of LSS with increased temperature. The analysis of the results can be summarized as follows:

- Integration of the phenoxy coupling layer in the composite structure was carried out in one shot during the used compression molding process, which means no additional energy or process step is required. The integration of the film was robust and, even in the compression-molding process with high-pressure gradients, no movement of the film occurred. Furthermore, no overflow of resin on the welding surface was visible and no visible shrinkage of the film during heating to the curing temperature of 140 °C appeared.

- In this study, it was presented that a strong connection between phenoxy and epoxy resin can be reached, even for very fast curing systems, with a curing time of 1 min at 140 °C.
- Lightweight joining technology with a robust ultrasonic welding process, a high average LSS of 24.4 MPa, and a standard deviation of 0.4 MPa have been achieved. The process time for welding was about 3 s, which is remarkable for the joining of a thermoset composite part. Based on the very linear storage modulus dependency on temperature of the phenoxy until the glass transition temperature (onset of 91 °C), the LSS of samples tested at 60 °C and 80 °C showed considerably high LSS results of 20.1 MPa and 18.3 MPa, respectively.
- Very short welding process times of 3 s for ultrasonic welding and 260 s for resistance welding with damage-tolerant joint design were reached, in comparison to state-of-the-art, epoxy-based adhesives. The welding process is surface-tolerant, which means no preparation is necessary. Furthermore, the proposed joining technology can be easily controlled and automated, and it is therefore adaptable for mass production.
- The use of thermoplastic as a joining material reduces the exposure of workers to chemicals (reactive adhesives). Furthermore, the thermoplastic coupling layer reduces the overall weight in comparison to mechanical fasteners and allows de-assembling possibilities. Due to all these factors, the overall environmental impact is reduced.

The performed study highlights the potential of welded hybrid joints based on a fast-curing epoxy composite with an integrated phenoxy coupling layer, as well as their application in the fields of general transport, wind energy, or sporting goods. In addition to the achieved results, there remain points that should be addressed in further studies, such as the long-term effects on fatigue, creep behavior, and humidity, to clarify the full potential of these weldable hybrid joints.

Author Contributions: Conceptualization, L.Z.; methodology, L.Z.; validation, L.Z.; formal analysis, C.B.; investigation, C.B.; resources, K.R.; writing—original draft preparation, L.Z. and C.B.; writing—review and editing, L.Z., C.B. and K.R.; visualization, C.B.; supervision, C.B. and K.R.; project administration, K.R. All authors have read and agreed to the published version of the manuscript.

Funding: This research received no external funding.

Institutional Review Board Statement: Not applicable.

Informed Consent Statement: Not applicable.

Data Availability Statement: The data presented in this study are available on request from the corresponding author.

Acknowledgments: We would like to thank Michel Lenoble and the company GabrielChem (now Huntsman Advanced Materials) for their involvement in the project, for their cooperation, and for the provision of materials.

Conflicts of Interest: The authors declare no conflict of interest.

References

1. Favaloro, M. Thermoplastic Composites in Aerospace–The Future Looks Bright. Available online: https://www.compositesworld.com/articles/thermoplastic-composites-in-aerospace-past-present-and-future (accessed on 15 December 2021).
2. Starke, J. Carbon Composites in Automotiv Structural Applications. Available online: https://cutt.ly/CteyLEP (accessed on 15 December 2021).
3. Meng, F.; Pickering, S.J.; Mckechnie, J. An Environmental Comparison of Carbon Fibre Composite Waste End-of-life Options. In Proceedings of the SAMPE Europe Conference, Southampton, UK, 11–13 September 2018.

4. Gardinger, G. New Horizons in Welding Thermoplastic Composite. Available online: https://www.compositesworld.com/articles/new-horizons-in-welding-thermoplastic-composites (accessed on 15 January 2022).
5. Wegmann, S.; Rytka, C.; Diaz-Rodenas, M.; Werlen, V.; Schneeberger, C.; Ermanni, P.; Caglar, B.; Gomez, C.; Michaud, V. A Life Cycle Analysis of Novel Lightweight Composite Processes: Reducing the Environmental Footprint of Automotive Structures. *J. Clean. Prod.* **2022**, *330*, 129808. [CrossRef]
6. Gardinger, G. Welding Thermoplastic Composites. Available online: https://www.compositesworld.com/articles/welding-thermoplastic-composites (accessed on 5 January 2022).
7. Shi, H.; Villegas, I.F.; Bersee, H.E.N. Strength and Failure Modes in Resistance Welded Thermoplastic Composite Joints: Effect of Fibre-Matrix Adhesion and Fibre Orientation. *Compos. Part A Appl. Sci. Manuf.* **2013**, *55*, 1–10. [CrossRef]
8. Zweifel, L.; Brunner, J.; Brauner, C.; Dransfeld, C. Development of a Resistance Welding Process for Thermoset Fiber Composite Components with Co-Cured Thermoplastic Boundary Layer. In Proceedings of the European Conference on Composite Materials, Athens, Greece, 25–28 June 2018.
9. Ageorges, C.; Ye, L. Resistance Welding of Thermosetting Composite/Thermoplastic Composite Joints. *Adv. Mater.* **2006**, *32*, 1603–1612. [CrossRef]
10. Stavrov, D.; Bersee, H. Resistance Welding of Thermoplastic Composites-An Overview. *Compos. Part A Appl. Sci. Manuf.* **2005**, *36*, 39–54. [CrossRef]
11. Ageorges, C.; Ye, L.; Hou, M. Advances in Fusion Bonding Techniques for Joining Thermoplastic Matrix Composites: A Review. *Compos. Part A Appl. Sci. Manuf.* **2001**, *32*, 839–857. [CrossRef]
12. Van Ingen, J.W.; Buitenhuis, A.; Van Wijngaarden, M.; Simmons, F. Development of the Gulfstream G650 Induction Welded Thermoplastic Elevators and Rudder. In Proceedings of the International SAMPE Symposium and Exhibition (Proceedings), Seattle, WA, Canada, 17–20 May 2010.
13. Villegas, I.F.; van Moorleghem, R. Ultrasonic Welding of Carbon/Epoxy and Carbon/PEEK Composites through a PEI Thermoplastic Coupling Layer. *Compos. Part A Appl. Sci. Manuf.* **2018**, *109*, 75–83. [CrossRef]
14. Palardy, G.; Villegas, I.F. Smart Ultrasonic Welding of Thermoplastic Composites. In Proceedings of the the American Society for Composites-31st Technical Conference, ASC 2016, Williamsburg, VA, USA, 19–22 September 2016.
15. Jongbloed, B.; Teuwen, J.; Palardy, G.; Fernandez Villegas, I.; Benedictus, R. Continuous Ultrasonic Welding of Thermoplastic Composites: Enhancing the Weld Uniformity by Changing the Energy Director. *J. Compos. Mater.* **2020**, *54*, 2023–2035. [CrossRef]
16. Troughton, M.J. *Handbook of Plastics Joining: A Practical Guide*, 2nd ed.; Elsevier: Amsterdam, The Netherlands, 2008; ISBN 9780815519768.
17. Offringa, A.; Myers, D.; Buitenhuis, A. Redesigned A340–500/600 Fixed Wing Leading Edge (J-Nose) in Thermoplastics. In Proceedings of the 22nd International SAMPE Europe Conference, Paris, France, 26–28 March 2001; p. 331.
18. Gardinger, G. Thermoplastic Composites Gain Leading Edge on the A380. Available online: https://www.compositesworld.com/articles/thermoplastic-composites-gain-leading-edge-on-the-a380 (accessed on 5 January 2022).
19. Thermoplastic Rudder and Elevator in G650 Empennage. Available online: https://www.gknaerospace.com/en/our-technology/2017/thermoplastic-rudder-and-elevator-in-g650-empennage/ (accessed on 5 January 2022).
20. Veldman, S.L.; Kortbeek, P.; Wölcken, P.C.; Kos, H.; Villegas, J. Development of a Multifunctional Fuselage Demonstrator. In Proceedings of the Aerospace Europe Conference, Bordeaux, France, 25–28 February 2020.
21. Omairey, S.L.; Sampethai, S.; Hans, L.; Worrall, C.; Lewis, S.; Negro, D.; Sattar, T.; Ferrera, E.; Blanco, E.; Wighton, J.; et al. Development of innovative automated solutions for the assembly of multifunctional thermoplastic composite fuselage. *Int. J. Adv. Manuf. Technol.* **2021**, *117*, 480–491. [CrossRef]
22. Zweifel, L.; Zhilyaev, I.; Brauner, C.; Rheme, M.; Eckhard, G.; Bersier, V.; Glavaški, S.; Pfeiffer, R. Experimental and Numerical Development on Multi-Material Joining Technology for Sandwich-Structured Composite Materials. *Materials* **2021**, *14*, 6005. [CrossRef]
23. MM-Welding®–Smart Joining of Lightweight Materials. Available online: https://www.bossard.com/ch-en/product-solutions/brands/welding-technology/mm-welding/ (accessed on 19 January 2022).
24. Engelschall, M.; Larsen, L.; Fischer, F.; Kupke, M. Robot-Based Continuous Ultrasonic Welding for Automated Production of Aerospace Structures. In Proceedings of the SAMPE Europe Conference, Nantes, France, 17–19 September 2019.
25. Don, R.C.; Gillespie, J.W., Jr.; McKnight, S.H. Bonding Techniques for High Performance Thermoplastic Compositions. U.S. Patent No. 5643390, 1 July 1997.
26. Brauner, C.; Nakouzi, S.; Zweifel, L.; Tresch, J. Co-Curing Behaviour of Thermoset Composites with a Thermoplastic Boundary Layer for Welding Purposes. *Compos. Adv. Mater.* **2020**, *29*, 1–9. [CrossRef]
27. Zweifel, L.; Brauner, C. Investigation of the Interphase Mechanisms and Welding Behaviour of Fast-Curing Epoxy Based Composites with Co-Cured Thermoplastic Boundary Layers. *Compos. Part A Appl. Sci. Manuf.* **2020**, *139*, 106120. [CrossRef]
28. Lestriez, B.; Chapel, J.P.; Gérard, J.F. Gradient Interphase between Reactive Epoxy and Glassy Thermoplastic from Dissolution Process, Reaction Kinetics, and Phase Separation Thermodynamics. *Macromolecules* **2001**, *34*, 1204–1213. [CrossRef]
29. Lipatov, Y.S.; Alekseeva, T.T. Phase-Separated Interpenetrating Polymer Networks. In *Phase-Separated Interpenetrating Polymer Networks*; Springer: Berlin/Heidelberg, Germany, 2007; pp. 1–227.

30. Teuwen, J.J.E.; Asquier, J.; Inderkum, P.; Masania, K.; Brauner, C.; Villegas, I.F.; Dransfeld, C. Gradient Interphases between High Tg Epoxy and Polyetherimide for Advanced Joining Processes. In Proceedings of the ECCM-18 Conference, Athens, Greece, 25–28 June 2018.
31. Farooq, U.; Heuer, S.; Teuwen, J.; Dransfeld, C. Effect of a Dwell Stage in the Cure Cycle on the Interphase Formation in a Poly(ether imide)/High Tg Epoxy System. *ACS Appl. Polym. Mater.* **2021**, *3*, 6111–6119. [CrossRef]
32. Van Krevelen, D.W.W. *Properties of Polymers*; Elsevier: Amsterdam, The Netherlands, 2009; ISBN 9780444828774.
33. Tsiangou, E.; Teixeira de Freitas, S.; Fernandez Villegas, I.; Benedictus, R. Investigation on Energy Director-Less Ultrasonic Welding of Polyetherimide (PEI)-to Epoxy-Based Composites. *Compos. Part B Eng.* **2019**, *173*, 107014. [CrossRef]
34. Bruckbauer, P. *Struktur-Eigenschafts-Beziehungen von Interphasen Zwischen Epoxidharz und Thermoplastischen Funktionsschichten für Faserverbundwerkstoffe*; TU München: München, Germany, 2018.
35. Zweifel, L.; Brauner, C.; Teuwen, J.; Dransfeld, C. In Situ Characterization of the Reaction-Diffusion Behavior during the Gradient Interphase Formation of Polyetherimide with a High-Temperature Epoxy System. *Polymers* **2022**, *14*, 435. [CrossRef]
36. SikaPower®-1200. Available online: https://industry.sika.com/en/home/renewable-energies/wind-energy/blade-manufacturing/surface-finishingandrepair/sikapower-1200.html (accessed on 15 January 2022).
37. 3M™ Structural Adhesive SA9820. Available online: https://www.3m.com/3M/en_US/p/d/b40066524/ (accessed on 18 December 2021).
38. ARALDITE® 2012. Available online: https://www.huntsman.com/products/araldite2000/araldite-2012 (accessed on 18 December 2021).
39. SikaPower®-1277. Available online: https://industry.sika.com/en/home/transportation/structural-adhesives/metal-adhesives/sikapower-1277.html (accessed on 15 January 2022).
40. Phenoxy PKHH PELLETS. Available online: https://www.gabrielchem.com/product/pkhh-pellets/ (accessed on 2 December 2021).
41. Dynamic Fluid Compression Molding. Available online: https://www.compositesworld.com/cdn/cms/DFCMOverview.pdf (accessed on 5 January 2022).
42. Huntsman Advanced Materials Develops a New Compression Molding Process. Available online: https://www.huntsman.com/about/advanced-materials/news/detail/12600/huntsman-advanced-materials-develops-a-new-compression (accessed on 7 January 2022).
43. Alif, N.; Carlsson, L.A.; Boogh, L. The Effect of Weave Pattern and Crack Propagation Direction on Mode I Delamination Resistance of Woven Glass and Carbon Composites. *Compos. Part B Eng.* **1998**, *29*, 603–611. [CrossRef]
44. Villegas, I.F. Ultrasonic Welding of Thermoplastic Composites. *Front. Mater.* **2019**, *6*, 291. [CrossRef]
45. Palardy, G.; Villegas, I.F. On the Effect of Flat Energy Directors Thickness on Heat Generation during Ultrasonic Welding of Thermoplastic Composites. *Compos. Interfaces* **2017**, *24*, 203–214. [CrossRef]
46. Villegas, I.F.; Palardy, G. Ultrasonic Welding of CF/PPS Composites with Integrated Triangular Energy Directors: Melting, Flow and Weld Strength Development. *Compos. Interfaces* **2017**, *24*, 515–528. [CrossRef]
47. Fernandez Villegas, I.; Valle Grande, B.; Bersee, H.E.N.; Benedictus, R. A Comparative Evaluation between Flat and Traditional Energy Directors for Ultrasonic Welding of CF/PPS Thermoplastic Composites. *Compos. Interfaces* **2015**, *22*, 717–729. [CrossRef]
48. Da Silva, L.F.M.; Öchsner, A.; Adams, R.D. *Handbook of Adhesion Technology*, 2nd ed.; Springer: Berlin/Heidelberg, Germany, 2018; Volumes 1–2.
49. Kupski, J. *Novel Adherend Laminate Designs for Composite Bonded Joints*; Delft University of Technology: Delft, The Netherlands, 2020.
50. Rajagopalan, G.; Gillespie, J.W.; McKnight, S.H. Diffusion of Reacting Epoxy and Amine Monomers in Polysulfone: A Diffusivity Model. *Polymer* **2000**, *41*, 7723–7733. [CrossRef]
51. Voleppe, Q.; Pardoen, T.; Bailly, C. Interdiffusion and Phase Separation upon Curing in Thermoset-Thermoplastic Interphases Unravelled by the Characterization of Partially Cured Systems. *Polymer* **2016**, *106*, 120–127. [CrossRef]
52. Villegas, I.F. Strength Development Versus Process Data in Ultrasonic Welding of Thermoplastic Composites with Flat Energy Directors and its Application to the Definition of Optimum Processing Parameters. *Compos. Part A Appl. Sci. Manuf.* **2014**, *65*, 27–37. [CrossRef]
53. Schollerer, M.J.; Kosmann, J.; Völkerink, O.; Holzhüter, D.; Hühne, C. Surface Toughening–A Concept to Decrease Stress Peaks in Bonded Joints. *J. Adhes.* **2019**, *95*, 495–514. [CrossRef]
54. Brauner, C.; Küng, M.; Arslan, D.; Maurer, C. Fused Filament Fabrication Based on Polyhydroxy Ether (Phenoxy) Polymers and Related Properties. *Polymers* **2021**, *13*, 1549. [CrossRef] [PubMed]

materials

Article

Facile Surface Depolymerization Promotes the Welding of Hard Epoxy Vitrimer

Le An * and Wenzhe Zhao

State Key Laboratory for Strength and Vibration of Mechanical Structures, Department of Engineering Mechanics, Xi'an Jiaotong University, Xi'an 710049, China; wenzhe0301@stu.xjtu.edu.cn
* Correspondence: anle2018@mail.xjtu.edu.cn

Abstract: Welding via bond exchange reactions has provided advances in obtaining high-quality joining performance. However, the reported welding method requires a relatively high press force, and challenges are still encountered in welding hard vitrimer. In this work, a facile surface depolymerization strategy was introduced to weld high-performance epoxy vitrimer. The vitrimers were firstly dissolved into ethylene glycol for depolymerization based on the solvent-assisted bond exchange reactions. Then, the depolymerized vitrimers were welded under heat and press force. The effect of the depolymerizing time, welding pressure, welding temperature and welding time on the welding strength were further investigated. It was found that there were optimal values for the depolymerizing time, welding pressure, and welding temperature, respectively, for the welding strength, while the welding strength increased with increasing welding time. Through facile surface degradation, the welding pressure was highly reduced, while the welding strength was increased. With surface depolymerization, the welding strength was 1.55-times higher, but the magnitude of press force was 1/1000-times than that with no surface depolymerization. It is elucidative that surface depolymerization can be used to weld hard vitrimer composites alongside reducing the press force effectively.

Keywords: epoxy vitrimer; surface depolymerization; welding; bond exchange reactions

Citation: An, L.; Zhao, W. Facile Surface Depolymerization Promotes the Welding of Hard Epoxy Vitrimer. *Materials* **2022**, *15*, 4488. https://doi.org/10.3390/ma15134488

Academic Editors: Patricia Krawczak, Chung-Hae Park and André Chateau Akué Asséko

Received: 6 May 2022
Accepted: 23 June 2022
Published: 25 June 2022

Copyright: © 2022 by the authors. Licensee MDPI, Basel, Switzerland. This article is an open access article distributed under the terms and conditions of the Creative Commons Attribution (CC BY) license (https://creativecommons.org/licenses/by/4.0/).

1. Introduction

Vitrimer is a novel kind of thermoset, where dynamic covalent bonds introduced into the cross-linked network can break, re-connect or exchange under external stimuli [1–5]. This type of polymer not only behaves as a conventional thermoset in terms of chemo-mechanical performance, but also enables cross-linked polymers to be reprocessed and recycled in a solid state. In the past decade, different vitrimers have been developed based on different exchangeable bonds, e.g., ester bond, disulfide and hindered urea bond [5–9]. Among them, epoxy-based transesterification is the most studied vitrimer chemistry due to its easy mass production [10–13]. Epoxy vitrimer is readily developed by adding an appropriate catalyst (such as zinc acetylacetonate ($Zn(acac)_2$) [11], and triazobicyclodecene (TBD) [14], etc.) into classical formulations of epoxy, which have been widely used in the fields of aerospace, electrical engineering, and coatings [15]. However, one of the main limitations of the epoxy vitrimer is that the catalysts are toxic and corrosive, which may cause corrosion damage to materials [16]. To date, vitrimers-based polymer blends have not yet been studied systemically.

Epoxy vitrimer is intrinsically composed of ester bonds and hydroxyl groups, rendering bond exchange reactions (BERs) attainable, which makes it possible for them to be welded from a bulk state or ground powder state. Yu et al. [17–19] used a constant press force of kPa to weld and recycle soft epoxy vitrimer, and the recycled material restored the mechanical properties to the level of the fresh material. For hard epoxy vitrimer, the welding pressure increased to MPa [20]. Luzuriaga et al. ground the epoxy vitrimer composites into powder and obtained the second-generation composites under 20 MPa pressure [1] Chabert et al. further proved the repeated welding ability of epoxy vitrimer composites in comparison with traditional thermosetting epoxy composites [21]. Compared with soft materials, chain diffusion in hard vitrimer poses significant difficulties and challenges due to the movement resistance of the densely cross-linked network. A large press force should be applied to make sure the welding surfaces are in full contact for welding. However, an excessive press force might wreak damage on the material due to its ductile nature.

Additionally, attempts have been made to lower the press force for welding epoxy vitrimer, where BERs were triggered between hydroxyl groups in alcohol solvent (e.g., EG, 2E1H) and esters in the epoxy network [22–24]. This type of BER containing a solvent is called a solvent-assisted BER and provides an efficient and mild way to recycle the polymer matrix, as well as the carbon fibers, with almost 100% recovered mechanical properties in a carbon fiber composite [23]. Based on this, Shi et al. [25] proposed a solvent-assisted welding strategy for the first time to realize welding under tiny press force, where, before welding, the surface of the soft epoxy vitrimer is depolymerized by an alcohol solvent. The essential degradation step prolonged the welding time; nevertheless, it provided an easy and effective bonding strategy to assemble a composite under smaller pressure or without pressure. To meet the increasing needs of industrial applications, hard materials with high cross-linking density are in great demand. To the best of our knowledge, there are, up to date, no publications on the welding of high-performance epoxy vitrimer.

In this paper, solvent-assisted BERs, which have been used for recycling rather than welding, are used to weld hard epoxy vitrimer. Ethylene glycol (EG), which offers hydroxyl groups, reacts with ester bonds of epoxy vitrimer reversibly. The vitrimer is firstly depolymerized and then forms a gel layer, in which the depolymerized polymer can diffuse across the welding interface easily. This will highly reduce the welding pressure. On the other hand, the depolymerized polymer can be repolymerized in the welding region, resulting in strong covalent bonds at the welding interface. This paper is organized as follows. In Section 1, we introduce the vitrimer and summarize the welding. In Section 2, the mechanism of surface depolymerizing and welding is demonstrated. In Section 3, we describe the welding procedures and characterizations. In Section 4, we present the results of the welding behavior of epoxy vitrimer. The focus of the study is to examine the suitable depolymerizing and welding conditions to achieve high welding strength alongside reducing the welding pressure. The concluding remarks are presented in Section 5.

2. Depolymerizing and Welding Mechanism

The polymer network of vitrimer can be depolymerized into small segments, and further be repolymerized thanks to the reversibility of solvent-assisted BERs [22–24]. Accordingly, the depolymerizing and welding of vitrimer are shown in Figure 1. Vitrimer is firstly immersed into an excess amount of solvent, where the solvent molecules effectively diffuse into the polymer and swell the fresh network. Then, the swelled network is gradually depolymerized as the depolymerized network, where a long polymer chain is broken into some small segments, as shown in Figure 1a. The depolymerized network is taken out of the solvent. As shown in Figure 1b, as two pieces of the depolymerized network touch each other, the small segments diffuse across the welding interface and repolymerize. Meanwhile, the re-generated solvent molecules escape from the repolymerized network and gradually evaporate. Compared with long polymer chains in the fresh network, the small segments in the depolymerized network are prone to diffusion [25]. Moreover, the depolymerized network is in a viscous gel state [26], guaranteeing good contact. Therefore, the facile surface depolymerization generally benefits chain diffusion, promoting the welding of vitrimer under small press force.

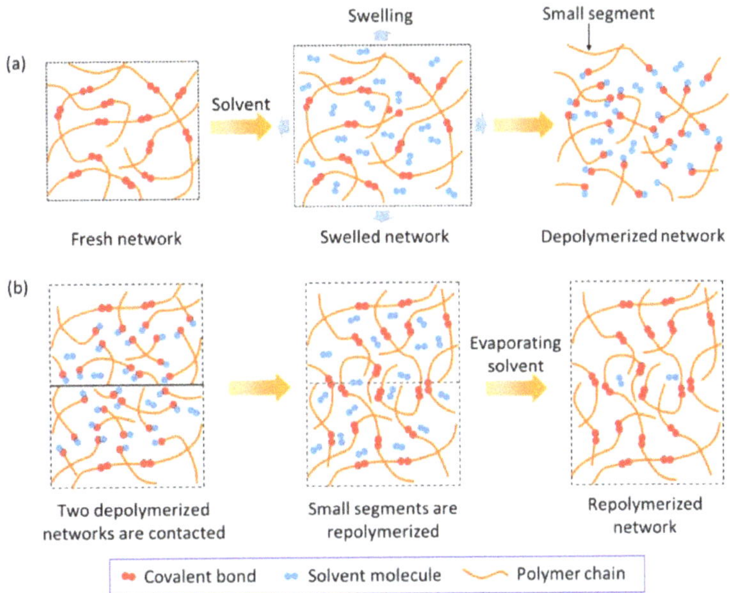

Figure 1. Schematic representation of depolymerizing and welding for vitrimer. (**a**) Original network is depolymerized into small segments. (**b**) Two depolymerized networks are repolymerized for welding.

3. Experimental

3.1. Materials

Diglycidyl ether of bisphenol A (DGEBA, MW: 340.4 g/mol, Sigma Aldrich, St. Louis, MO, USA), glutaric anhydride (GA, MW: 114.10 g/mol, Macklin, Shanghai, China), zinc acetylacetonate hydrate ($Zn(ac)_2$, Sigma Aldrich, St. Louis, MO, USA), and ethylene glycol (EG, AR grade, Macklin, Shanghai, China).

3.2. Preparation of Hard Epoxy Vitrimer

Hard epoxy vitrimer was prepared with the following synthesis procedures: DGEBA and GA in solid states were heated at 130 °C until melted. Then, 5 mol % $Zn(ac)_2$ to the epoxy groups was added into melted DGEBA and stirred powerfully at 130 °C for 10 min. Then, the melted GA was poured into the mixture with a stoichiometric ratio of epoxy groups and acyl groups of 1:1. The above mixture was poured into home-made molds to cure at 140 °C for 12 h after undergoing vacuum degassing at 100 °C for 15 min. The mechanical and thermal properties of the epoxy vitrimer are presented in Appendix A.

3.3. Dissolving Experiment

An appropriate depolymerization condition for welding was roughly obtained by a dissolving experiment for epoxy vitrimer. In our previous work [16], the dissolving temperature for epoxy vitrimer was located at 140–200 °C, which was around the boiling point of EG (197.3 °C) and far less than the pyrolysis temperature of the epoxy vitrimer (Table A1). Herein the dissolving temperature was 180 °C. The mass ratio of epoxy vitrimer to EG was 1:25. A cubic epoxy vitrimer sample with the length of 10 mm was immersed into EG solvent, then placed in an oven and heated under 180 °C. At different time intervals, the sample was taken out and weighed immediately to record the mass of the residual vitrimer.

3.4. Welding Experiments

The solvent-assisted welding process is shown in Figure 2. Two separated vitrimer samples (length 25 mm, width 15 mm, and thickness 8 mm) (Figure 2I) were put into

EG solvent at a depolymerizing temperature of T_1 for a depolymerizing time of t_1 in a vacuum oven (Figure 2II). Then, two depolymerized vitrimer were connected under a welding pressure of P at a welding temperature of T_2 for a welding time of t_2, where the area of the butt joint was 15 mm × 8 mm (Figure 2III). All of the depolymerizing as well as welding conditions are summarized in Group 1–4 in Table 1. The welded vitrimer with a butt-joint was cut along the *y*-*z* plane to prepare stretching specimens with a thickness of ~2 mm (Figure 2IV). During welding, the welding pressure and welding temperature were provided by a hot-pressing apparatus (Figure 3), in which the heating rate and cooling rate were 20 °C/min and 10 °C/min, respectively. Moreover, a steel mold was fabricated to prevent the vitrimer sample from buckling during the welding process.

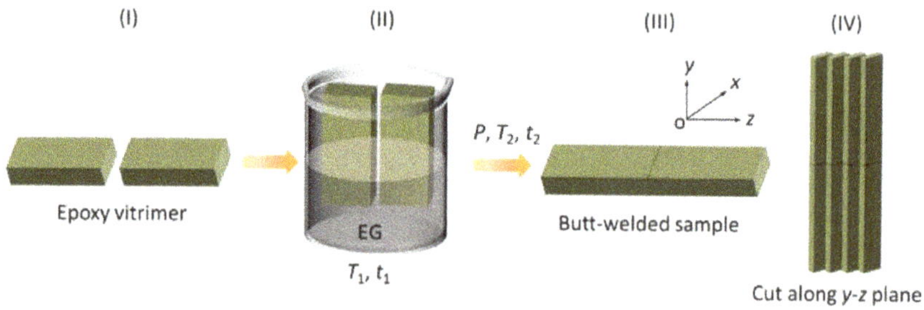

Figure 2. Schematic of the solvent-assisted welding process of epoxy vitrimer: (**I**) original epoxy vitrimer, (**II**) immersing into EG, (**III**) butt-welded vitrimer, and (**IV**) cut specimens for stretching.

Table 1. Depolymerizing and welding conditions for epoxy vitrimer.

Group	Depolymerizing Time t_1 (min)	Depolymerizing Temperature T_1 (°C)	Welding Pressure P (kPa)	Welding Temperature T_2 (°C)	Welding Time t_2 (h)
Group 1	20, 30, 40, 50	180	20	180	12
Group 2	40	180	0, 10, 20, 30	180	12
Group 3	40	180	20	140, 160, 180, 200	12
Group 4	40	180	20	180	4, 6, 8, 10, 12
Group 5	/	/	1000, 5000, 10,000, 20,000	180	12

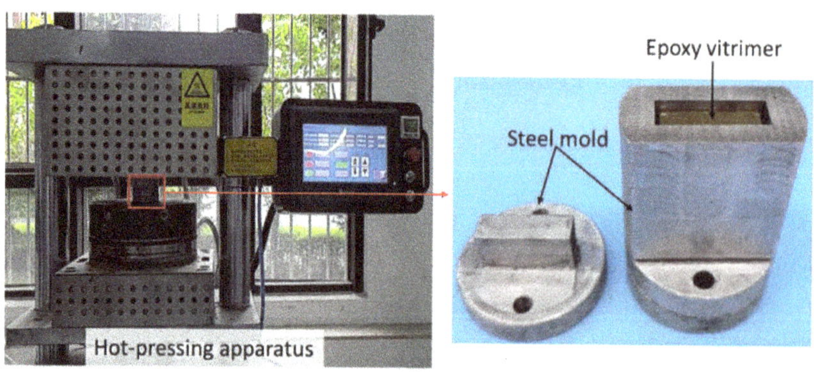

Figure 3. Experimental setup for welding.

The dimension of the vitrimer for hot-press welding was the same as those for solvent-assisted welding. The hot-press welding process was carried out as follows. The epoxy vitrimer samples were cleaned with ethyl alcohol and then dried at 100 °C for 10 min to evaporate the ethyl alcohol. Then, the vitrimer samples were contacted using the hot-pressing apparatus and the steel mold (Figure 3). The welding pressure, welding temperature, and welding time are shown in Group 5 in Table 1.

3.5. Characterization

The butt-welded vitrimer was tested using a universal material testing machine (MTS Criterion C45.105, Eden Prairie, MN, USA) with a tensile rate of 1 mm/min. The force and crosshead displacement were recorded. The strain was calculated as the displacement divided by the initial distance of the crosshead. The stress was calculated as the force divided by the welding area in the undeformed state. Thus, the welding strength of epoxy specimens was calculated by:

$$\tau = \frac{F_{max}}{S} \quad (1)$$

where F_{max} was the maximum load and, S was the welding area in the undeformed state. At least three specimens were tested for each case.

The surface morphologies of the welded region were investigated by a 3D microscope system (LY-WN-YH System, Liyang Co., Ltd., Chengdu, Sichuan, China).

4. Results and Discussions

4.1. The Influence of Depolymerizing Time

The depolymerizing time for welding was determined by the dissolving experiment of cubic epoxy vitrimer into EG solvent at 180 °C, as shown in Figure 4a. During dissolving, the EG solvent provided abundant hydroxyl groups to break the esters on the polymer network, leading to depolymerization of the vitrimer. The dissolving process involved three processes [16], e.g., diffusion of EG molecules into the network, the cleavage of the polymer chains due to solvent-assisted BERs between the hydroxyl groups of the EG solvent and esters of the network, and the diffusion of small segments into the EG solvent. Figure 4b shows that the cubic shape of the vitrimer remains unchanged during the dissolving process, indicating that the dissolution of the vitrimer is a typical surface erosion mode [22,26].

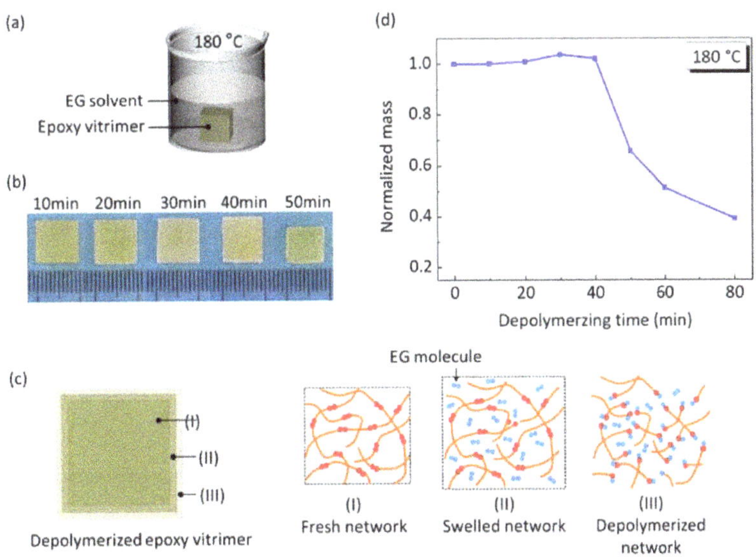

Figure 4. Depolymerization of epoxy vitrimer being immersed into EG at 180 °C. (**a**) Dissolving experiment of the vitrimer. (**b**) The appearance and size evolution of the depolymerized vitrimer. (**c**) Schematic representation of the depolymerized vitrimer. (**d**) Normalized mass of depolymerized vitrimer as a function of depolymerizing time.

After partially dissolving, the depolymerized vitrimer was assumed to contain three parts: the fresh network, the swelled network, and the depolymerized network, as shown in Figure 4c. The swelled network and the depolymerized network are much smaller than the fresh network. In the swelled network, a small amount of EG molecules are absorbed but do not break the network. In the depolymerized network, the network is depolymerized into small segments due to solvent-assisted BERs. Therefore, the mass of the depolymerized vitrimer is the total mass containing absorbed EG molecules in the swelled network and depolymerized network. Here, the mass of the depolymerized vitrimer is normalized by the initial mass, as shown in Figure 4d. In the first 30 min, the mass of the depolymerized vitrimer increases slightly due to the diffusion of EG molecules in the swelled network; however, the depolymerized network is very small. When the dissolving time increases from 30 min to 40 min, the absorbed EG continues to increase but the depolymerized mass is still small. The mass of the depolymerized vitrimer decreases sharply after dissolving for 40 min, in which there is an obvious material loss. In conclusion, the surface depolymerization has dual characteristics for welding. On the one hand, it contributes to breaking the fresh network, which will promote the followed welding process. On the other hand, it gives rise to vitrimer loss, which will be detrimental to the vitrimer integrity.

To quantitatively provide criteria for optimizing the depolymerizing time, we welded the depolymerized vitrimer and compared their welding strength. At the beginning of the welding, two pieces of the depolymerized vitrimer specimens are connected to each other (Figure 5a). The depolymerized network is diffused across the interface. After being welded, the absorbed EG molecules in the swelled network are evaporated, as shown in Figure 5b. Meanwhile, the generated solvent molecules evaporate from the interface. Thus, the depolymerizing conditions affect the degree of depolymerization, further affecting the welding strength. For a given depolymerizing temperature, the welding strength is only related to the depolymerizing time.

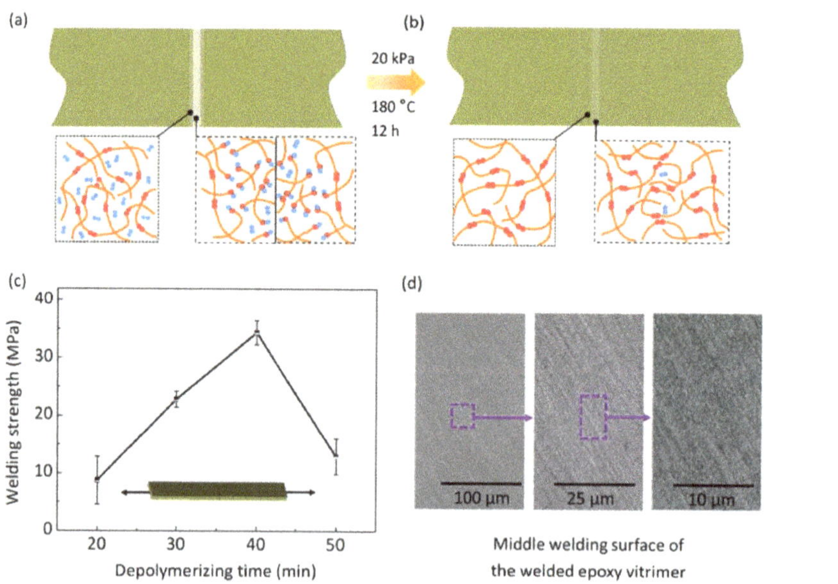

Figure 5. Effect of depolymerizing time on the welding performance of epoxy vitrimers. The vitrimer are welded under 20 kPa at 180 °C for 12 h. Schematic representation of the depolymerized vitrimers (**a**) before welding and (**b**) after welding. (**c**) The welding strength of the butt-welded vitrimer versus depolymerizing time. Each data point represents the mean and stands deviation of three measurements. (**d**) The microscopical images of welded region with depolymerizing time of 40 min.

After being depolymerized for 20 min, 30 min, 40 min, and 50 min, respectively, the depolymerized vitrimer were welded at 180 °C under 20 kPa for 12 h (Group 1 in Table 1). The tensile behavior of the butt-welded specimens is shown in Figure 5c. The greatest welding strength is about 35 MPa for a depolymerizing time of 40 min (54% of the fresh vitrimer), while the smallest welding strength is about 8 MPa for a depolymerizing time of only 20 min. The welding strength decreases to 12 MPa as the depolymerizing time increases to 50 min, which is lower than that for a depolymerizing time of 30 min and 40 min. The trend that the welding strength increases at first and then declines with depolymerizing time can be explained. On the one hand, the content of EG absorbed in the polymer network is very limited within 20 min, which is not beneficial to repolymerization for the subsequent welding process. On the other hand, the surface of the vitrimer is over-depolymerized for 50 min, making repolymerization difficult due to the imperfect surface contact. The vitrimer under a depolymerizing time of 40 min possesses the largest swelling content of EG, which improves depolymerization, while no severe fracturing of the vitrimer occurs.

Figure 5d shows the microscopic images of the welded region with a depolymerizing time of 40 min and a welding time of 12 h. It was found that there is no obvious welding interface from the middle surface view, which reflects efficient welding with zero-interface. However, the failure of the welded specimen substantially originates from the welding interface because incomplete repolymerization with insufficient welding conditions was provided.

4.2. The Influences of Welding Conditions

The depolymerized network will be re-polymerized due to the solvent-assisted BERs, which can be reversed using the depolymerizing process [23–25]. According to our previous work for recycling epoxy, visible hydroxyl groups unavoidably exist in the repolymerized network and the residual EG molecules substantially reduce the degree of the repolymerized network [16]. Thus, the welding quality highly depends on the welding conditions, e.g., welding pressure, welding temperature and welding time. Herein we study how the above welding conditions affect the welding strength, and the corresponding results are shown in Figure 6. Before welding, the vitrimer samples are depolymerized at 180 °C for 40 min.

Figure 6a, d show the influence of welding pressure on stress–strain curves and the welding strength, respectively. The welding temperature is 180 °C and the welding time is 12 h (Group 2 in Table 1). The welding strength is 17.2 MPa for the case with a lack of welding pressure. As the welding pressure increases to 20 kPa, the welding strength increases to 34 MPa, indicating that a higher press force increases the contact area and further enhances the welding strength. However, the welding strength decreases slightly as the welding pressure increases to 30 MPa. This is possibly because excess press force squeezes out the depolymerized network and further prevents depolymerized chain diffusion at the welding interface. Therefore, increasing the press force does not significantly promote the welding efficiency, which is completely different from the reported results, indicating that press force accelerates welding without surface depolymerization [19,20].

Figure 6b,e show the influence of welding temperature on stress–strain curves and the welding strength, respectively. The welding pressure is 20 kPa and the welding time is 12 h (Group 3 in Table 1). The welding strength is only 4 MPa when the welding temperature is 140 °C, and reaches 35 MPa as the temperature approaches 180 °C. This is because higher temperature promotes solvent-assisted BERs and, likewise, EG evaporation simultaneously. The degree of repolymerized network enlarges while the residual EG lessens, resulting in improved welding strength. As the welding temperature increases to 200 °C, the welding strength is slightly lower than that at 180 °C. This is probably ascribed to the fact that the faster formation of the repolymerization network on the welding surface prevents the evaporation of enclosed EG molecules and further lowers the welding strength. Moreover, excessive welding temperature may lead to surface degradation due to oxidation.

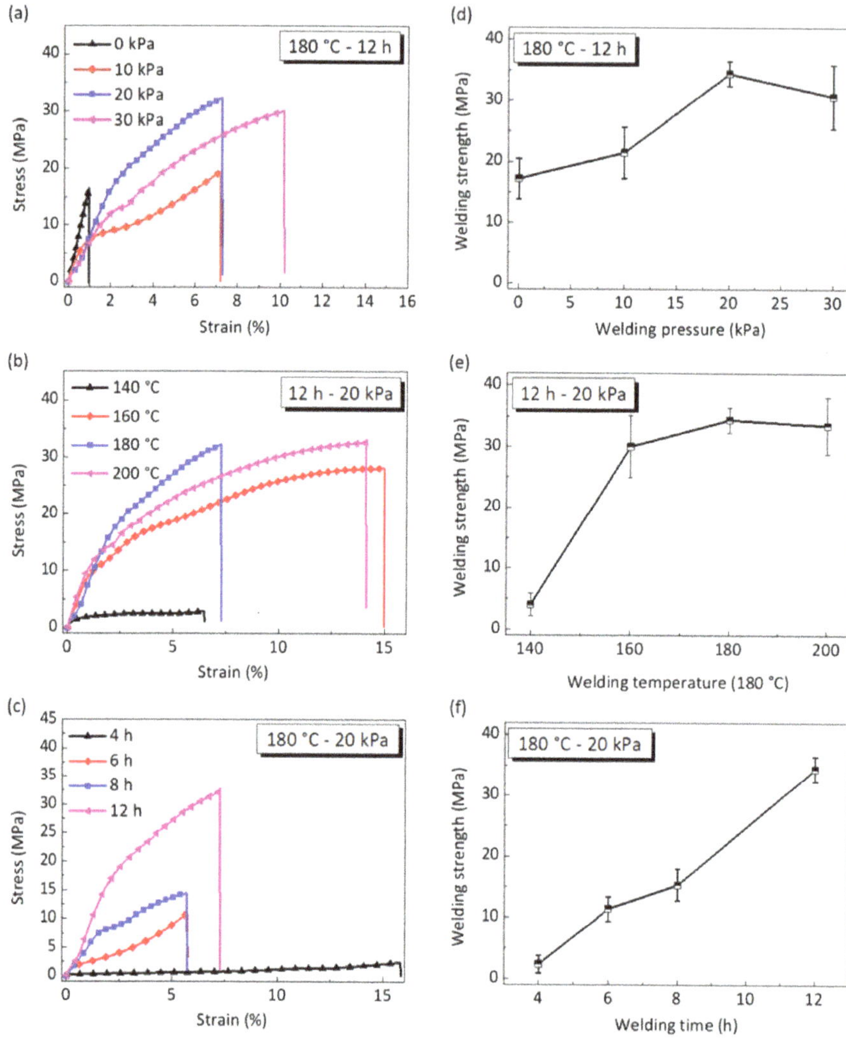

Figure 6. Effect of welding conditions on the mechanical behavior of butt-welded epoxy vitrimer. All of the epoxy vitrimers are depolymerized at 180 °C for 40 min before welding. Stress–strain curves under different (**a**) welding pressure, (**b**) welding temperature, and (**c**) welding time. The welding strength varies with (**e**) welding pressure, (**e**) welding temperature, and (**f**) welding time. Each data point represents the mean and stands deviation of three measurements.

Figure 6c, f show that the stress–strain curves and welding strength vary with a welding time of 4 h, 6 h, 8 h, and 12 h (Group 4 in Table 1). The welding pressure and welding temperature are 20 kPa and 180 °C, respectively. It is seen that the welding strength gradually increases as the welding time increases. Sufficient welding time ensures the full completion of network repolymerization at the welding interface, which enables strong welding. Moreover, extending the welding time facilitates EG evaporation in the repolymerized network at the welding interface, which finally boosts the welding strength. Thus, we believe that a higher welding strength could be achieved if we further increase the welding time.

4.3. Comparison to Welding with no Surface Depolymerization

For comparison, the hard epoxy vitrimer are welded using hot-press welding without surface depolymerization, where the welding pressure is from 1 MPa to 20 MPa. Figure 7a shows the stress–strain curves of welded vitrimer as they are welded at 180 °C for 12 h (Group 5 in Table 1). All the curves are linear and have the same slope. Compared with Figure 6a,c, one can see that the break strain using hot-press welding is much smaller than that using solvent-assisted welding. In addition, the comparison of welding strength with and without surface depolymerization is shown in Figure 7b. Without surface depolymerization, welding under a large pressure of 10 MPa reaches to a welding strength of 16.8 MPa, close to the strength level of welding under zero pressure as the vitrimer is depolymerized. The maximum welding strength of 35 MPa is attained after surface depolymerization, 1.55-times higher than the welding at an extreme press force of 20 MPa without surface depolymerization, whereas the welding pressure of the former is 1/1000 of the latter. Therefore, we can conclude that the use of surface depolymerization encourages the fluidity of the polymer chains to diffuse readily, which further strikingly reduces the required press force for welding.

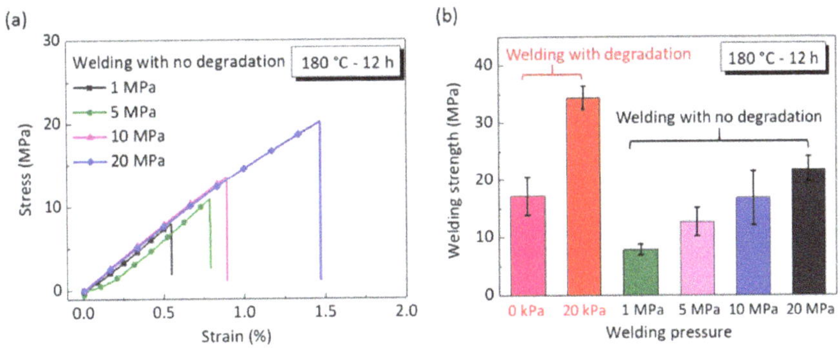

Figure 7. Comparison of the welding performance of epoxy vitrimer using hot-press welding and solvent-assisted welding. (**a**) Stress–strain curves of hot-press welding. (**b**) Welding strength of solvent-assisted welding and hot-press welding. The welding time and temperature are 12 h and 180 °C, respectively. Each data point represents the mean and stands deviation of three measurements.

Epoxy vitrimer can be intrinsically welded due to the self-contained dynamic covalent bonds [27]. In comparison with soft materials, the required welding pressure for vitrimers with high cross-linking densities should be larger as the polymers are too inactive to deform [17]. This work introduces a facile surface depolymerization strategy to reduce the welding pressure for welding high-performance epoxy vitrimer. The previous welding method has the following two advantages. On the one hand, the depolymerized materials on the surface of the vitrimer form a gel layer with a thickness of ~0.1 mm, which is prone to diffusing across the welding interface [25]. As a consequence, the welding pressure, which is used to make sure the welding surfaces come into contact, decreases from tens of megapascals to tens of kilopascals, as shown in Figure 7. On the other hand, large numbers of hydroxyl groups are introduced, which accelerate the bond exchange reaction and form covalent bonds at the welding interface, thus promoting strong welding (Figure 1). In view of the above-mentioned advantages, this surface degradation strategy has the potential for welding vitrimer composites to meet the increasing needs of industrial applications [27–29]. However, there are still some problems in the surface degradation for welding, such as the need for high welding temperatures, which are determined by the BERs and the boiling point of the EG. Therefore, it is worth further exploring solvents with low boiling points as well as more efficient BER chemistry for relatively low temperatures.

5. Conclusions

We have realized strong welding for hard epoxy vitrimer under small press force thanks to facile surface depolymerization. The vitrimer specimens are firstly dissolved into an excess amount of EG solvents for surface depolymerization and then welded as the depolymerized vitrimer specimen pieces touch each other. We first conducted dissolving experiments and welding experiments to determine the depolymerizing time of epoxy vitrimer, which is 40 min under 180 °C. Then, we investigated the influence of welding conditions on the welding strength of epoxy vitrimer. During the welding process, an appropriate welding temperature and long welding time are favored in attaining remarkable welding strength. Furthermore, increasing the welding pressure does not significantly promote the welding efficiency, which is completely different from the reported results of hot-press welding. Under the same conditions (180 °C, 12 h), the welding strength using surface depolymerization is 35 MPa under 20 kPa, while that using hot-pressing welding is only 21.9 MPa. That is, the welding strength using surface depolymerization is 1.55-times higher, but the magnitude of press force is 1/1000. This work provides an example of using surface depolymerization to develop welding methods for vitrimer composites. Nevertheless, these are comparatively high welding temperatures determined by the boiling point of EG. Therefore, it is worth further exploring solvents with low boiling points.

Author Contributions: Conceptualization, L.A.; methodology, L.A.; investigation, W.Z. and L.A.; writing—original draft preparation, W.Z.; writing—review and editing, L.A.; funding acquisition, L.A. All authors have read and agreed to the published version of the manuscript.

Funding: This research was funded by the National Natural Science Foundation of China (No. 12002257), China Postdoctoral Science Foundation (No. 2019M663676), and the Fundamental Research Funds for the Central Universities (No. xjh012019035).

Institutional Review Board Statement: Not applicable.

Informed Consent Statement: Not applicable.

Data Availability Statement: The data presented in this study are available on reasonable request from the corresponding author.

Conflicts of Interest: The authors declare no conflict of interest.

Appendix A. The Mechanical and Thermal Properties of Hard Epoxy Vitrimer

DSC measurements were conducted using a DSC1 differential scanning calorimeter (Mettler Toledo, Switzerland). The sample was heated from 0 °C to 250 °C at a rate of 10 °C/min. The DSC curve of epoxy vitrimer shows an obvious glass transition of 77.2 °C in the second heating curve, as shown in Figure A1a. The glass transition temperature is also list in Table A1.

Tensile tests of epoxy vitrimer were carried out using a universal material testing machine (MTS Criterion C45.105, Eden Prairie, MN, USA). The epoxy vitrimer samples were cut into a dog-bone shape and tested according to ASTM standard 638-14. The tensile stress–strain curve of epoxy vitrimer is shown in Figure A1b. The basic mechanical parameters are listed in Table A1.

Thermogravimetric analysis (TGA) was performed using a TGA instrument (Netzsch STA449F5, Germany). The experiments were performed within an air atmosphere at a ramp rate of 10 °C/min from 30 °C to 800 °C, where the gas flow rate was 50 mL/min. TGA profiles (Figure A1c) show that the decomposition temperatures exhibits two stages: the first one ranges from 330 °C to 482 °C due to pyrolysis and carbonization; while the second stage varies from 482 °C to 607 °C due to the decomposition of the carbonized products. All the thermal and mechanical properties are summarized in Table A1.

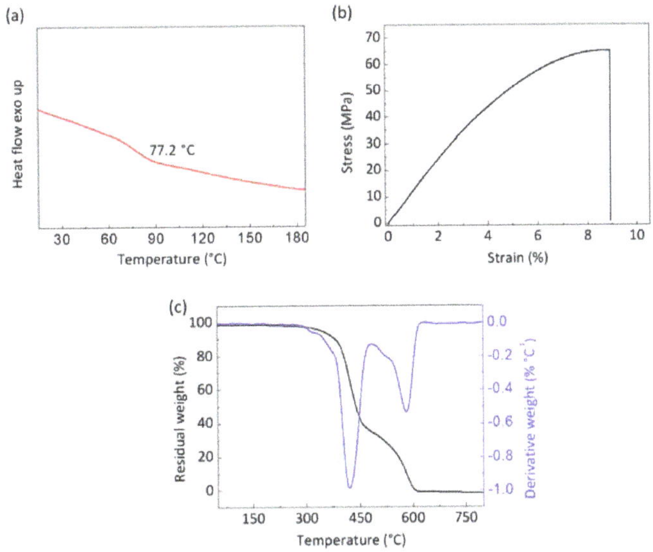

Figure A1. The mechanical and thermal properties of epoxy vitrimer. (**a**) DSC curve, (**b**) Stress–strain curve, (**c**) TGA and DTG curves.

Table A1. Thermal and mechanical properties of the cured epoxy vitrimer.

T_g/°C	Tensile Strength/MPa	Elastic Modulus/MPa	Elongation at Break/%	$T_{5\%}$/°C	$T_{10\%}$/°C	T_{max1}/°C	T_{max2}/°C
77.2	63.0 ± 3.4	1192.9 ± 34.5	8.6 ± 2.4	346	382	420	582

References

1. De Luzuriaga, A.R.; Martin, R.; Markaide, N.; Rekondo, A.; Cabañero, G.; Rodríguez, J.; Odriozola, I. Epoxy resin with exchangeable disulfide cross-links to obtain reprocessable, repairable and recyclable fiber-reinforced thermoset composites. *Mater. Horiz.* **2016**, *3*, 241–247. [CrossRef]
2. Van Zee, N.J.; Nicolaÿ, R. Vitrimers: Permanently crosslinked polymers with dynamic network topology. *Prog. Polym. Sci.* **2020**, *104*, 101233. [CrossRef]
3. Denissen, W.; Winne, J.M.; Du Prez, F.E. Vitrimers: Permanent organic networks with glass-like fluidity. *Chem. Sci.* **2016**, *7*, 30–38. [CrossRef] [PubMed]
4. Panagiotopoulos, C.; Porfyris, A.; Korres, D.; Vouyiouka, S. Solid-state polymerization as a vitrimerization tool starting from available thermoplastics: The effect of eeaction temperature. *Materials* **2020**, *14*, 9. [CrossRef]
5. Alabiso, W.; Schlögl, S. The impact of vitrimers on the industry of the future: Chemistry, properties and sustainable forward-looking applications. *Polymers* **2020**, *12*, 1660. [CrossRef]
6. Zheng, N.; Xu, Y.; Zhao, Q.; Xie, T. Dynamic covalent polymer networks: A molecular platform for designing functions beyond chemical recycling and self-healing. *Chem. Rev.* **2021**, *121*, 1716–1745. [CrossRef]
7. Scheutz, G.M.; Lessard, J.J.; Sims, M.B.; Sumerlin, B.S. Adaptable crosslinks in polymeric materials: Resolving the intersection of thermoplastics and thermosets. *J. Am. Chem. Soc.* **2019**, *141*, 16181–16196. [CrossRef]
8. Liu, T.; Zhao, B.; Zhang, J. Recent development of repairable, malleable and recyclable thermosetting polymers through dynamic transesterification. *Polymer* **2020**, *194*, 122392. [CrossRef]
9. Zhang, Y.; Ying, H.; Hart, K.R.; Wu, Y.; Hsu, A.J.; Coppola, A.M.; Kim, T.A.; Yang, K.; Sottos, N.R.; White, S.R.; et al. Malleable and recyclable poly(urea-urethane) thermosets bearing hindered urea bonds. *Adv. Mater.* **2016**, *28*, 7646–7651. [CrossRef]
10. Altuna, F.I.; Hoppe, C.E.; Williams, R.J.J. Epoxy vitrimers: The effect of transesterification reactions on the network structure. *Polymers* **2018**, *10*, 43. [CrossRef]
11. Capelot, M.; Montarnal, D.; Tournilhac, F.; Leibler, L. Metal-catalyzed transesterification for healing and assembling of thermosets. *J. Am. Chem. Soc.* **2012**, *134*, 7664–7667. [CrossRef] [PubMed]
12. Montarnal, D.; Capelot, M.; Tournilhac, F.; Leibler, L. Silica-like malleable materials from permanent organic networks. *Science* **2011**, *334*, 965–968. [CrossRef] [PubMed]

13. Ran, Y.; Zheng, L.J.; Zeng, J.B. Dynamic crosslinking: An efficient approach to fabricate epoxy vitrimer. *Materials* **2021**, *14*, 919 [CrossRef] [PubMed]
14. Liu, H.; Zhang, H.; Wang, H.; Huang, X.; Huang, G.; Wu, J. Weldable, malleable and programmable epoxy vitrimers with high mechanical properties and water insensitivity. *Chem. Eng. J.* **2019**, *368*, 61–70. [CrossRef]
15. Roudsari, G.M.; Mohanty, A.K.; Misra, M. Green approaches to engineer tough biobased epoxies: A review. *ACS Sustain. Chem. Eng.* **2017**, *5*, 9528–9541. [CrossRef]
16. Zhao, W.; An, L.; Wang, S. Recyclable high-performance epoxy-anhydride resins with DMP-30 as the catalyst of transesterification reactions. *Polymers* **2021**, *13*, 296. [CrossRef]
17. Yu, Y.; Storti, G.; Morbidelli, M. Kinetics of ring-opening polymerization ofl, l-lactide. *Ind. Eng. Chem. Res.* **2011**, *50*, 7927–7940. [CrossRef]
18. Yu, K.; Taynton, P.; Zhang, W.; Dunn, M.L.; Qi, H.J. Reprocessing and recycling of thermosetting polymers based on bond exchange reactions. *RSC Adv.* **2014**, *4*, 10108–10117. [CrossRef]
19. Yu, K.; Shi, Q.; Li, H.; Jabour, J.; Yang, H.; Dunn, M.L.; Wang, T.J.; Qi, H.J. Interfacial welding of dynamic covalent network polymers. *J. Mech. Phys. Solids* **2016**, *94*, 1–17. [CrossRef]
20. An, L.; Shi, Q.; Jin, C.; Zhao, W.; Wang, T.J. Chain diffusion based framework for modeling the welding of vitrimers. *J. Mech. Phys. Solids* **2022**, *164*, 104883. [CrossRef]
21. Chabert, E.; Vial, J.; Cauchois, J.-P.; Mihaluta, M.; Tournilhac, F. Multiple welding of long fiber epoxy vitrimer composites. *Soft Matter* **2016**, *12*, 4838–4845. [CrossRef] [PubMed]
22. Hamel, C.M.; Kuang, X.; Chen, K.; Qi, H.J. Reaction-diffusion model for thermosetting polymer dissolution through exchange reactions assisted by small-molecule solvents. *Macromolecules* **2019**, *52*, 3636–3645. [CrossRef]
23. Yu, K.; Shi, Q.; Dunn, M.L.; Wang, T.J.; Qi, H.J. Carbon fiber reinforced thermoset composite with near 100% recyclability. *Adv. Funct. Mater.* **2016**, *26*, 6098–6106. [CrossRef]
24. Mu, Q.; An, L.; Hu, Z.; Kuang, X. Fast and sustainable recycling of epoxy and composites using mixed solvents. *Polym. Degrad. Stab.* **2022**, *199*, 109895. [CrossRef]
25. Shi, Q.; Yu, K.; Dunn, M.L.; Wang, T.J.; Qi, H.J. Solvent assisted pressure-free surface welding and reprocessing of malleable epoxy polymers. *Macromolecules* **2016**, *49*, 5527–5537. [CrossRef]
26. Kuang, X.; Shi, Q.; Zhou, Y.; Zhao, Z.; Wang, T.J.; Qi, H.J. Dissolution of epoxy thermosets via mild alcoholysis: The mechanism and kinetics study. *RSC Adv.* **2018**, *8*, 1493–1502. [CrossRef]
27. Yang, Y.; Ding, X.; Urban, M.W. Chemical and physical aspects of self-healing materials. *Prog. Polym. Sci.* **2015**, *49*, 34–59. [CrossRef]
28. Ji, F.; Liu, X.; Sheng, D.; Yang, Y. Epoxy-vitrimer composites based on exchangeable aromatic disulfide bonds: Reprocessibility, adhesive, multi-shape memory effect. *Polymer* **2020**, *197*, 122514. [CrossRef]
29. Yang, Y.; Xu, Y.; Ji, Y.; Wei, Y. Functional epoxy vitrimers and composites. *Prog. Mater. Sci.* **2020**, *120*, 100710. [CrossRef]

materials

MDPI

Article

On the Hot-Plate Welding of Reactively Compatibilized Acrylic-Based Composites/Polyamide (PA)-12

Henri Perrin, Masoud Bodaghi *, Vincent Berthé, Sébastien Klein and Régis Vaudemont

Luxembourg Institute of Science and Technology (LIST), 5, rue Bommel, L-4940 Hautcharage, Luxembourg
* Correspondence: masoud.bodaghi@list.lu; Tel.: +352-2758884575

Abstract: Joining of dissimilar thermoplastics and their composites is a challenge for thermal welding techniques due to different melting points. Reactive welding with an auxiliary functional material can offer the clear opportunities to develop joining processes due to robustness to joining dissimilar thermoplastic polymers and their composites. The current study employed reactive compatibilization to offer the possibility of joining an acrylic-based glass fiber composite to polyamide (PA)-12 by applying a hot-tool welding technique. For this purpose, composite plates are fabricated by a typical vacuum infusion and thin layer thermoplastic films are formed by a thermostamping of PA12 granules. Subsequently, the reactive welding of the interposed PA12 sheet and Elium®-GMA-Glass composite is conducted by hot-plate welding. A glycidyl methacrylate (GMA) as a compatibilizing agent is copolymerized with methyl methacrylate Elium® resin. During the hot-tool welding process of dissimilar thermoplastic material, GMA can react with the polyamide end groups. The heat distribution at the Elium® GMA/PA-12 interface is responsible for obtaining a strong joint. This study focuses on the functionality of the compatibilizer on the welding of acrylic-based composites with polyamide (PA)-12 while varying the assembly temperature. The flatwise tensile test proved the effectiveness of GMA on the interface bounding. The excellent bounding incompatible polymers Elium® resin (PMMA) and PA12 was achieved at 200 °C.

Keywords: fusion bounding; Elium®; reactive welding

Citation: Perrin, H.; Bodaghi, M.; Berthé, V.; Klein, S.; Vaudemont, R. On the Hot-Plate Welding of Reactively Compatibilized Acrylic-Based Composites/ Polyamide (PA)-12. *Materials* **2023**, *16*, 691. https://doi.org/10.3390/ma16020691

Academic Editor: Tomasz Trzepieciński

Received: 27 December 2022
Revised: 4 January 2023
Accepted: 6 January 2023
Published: 10 January 2023

Copyright: © 2023 by the authors. Licensee MDPI, Basel, Switzerland. This article is an open access article distributed under the terms and conditions of the Creative Commons Attribution (CC BY) license (https://creativecommons.org/licenses/by/4.0/).

1. Introduction

The versatile application of multi material design consisting of dissimilar polymers and their composites is a possible response to the increasing demand of reducing weight and improving mechanical properties in automotive and aerospace industries. In practical applications, joining light thermoplastic and high-strength fiber reinforced thermoplastic composite is a widespread design [1,2].

Fusion bounding is a highly practical technique for joining fiber reinforced thermoplastic composites. Experience with the fusion bounding method has proven the possibility of joining dissimilar parts with eliminating stress concentration and rivets while staying cost effective. In this family of processes, after close contact of surfaces, inter-diffusion over a period allows joining the two polymeric parts or their composites above the certain temperature. Depending on the heating mechanisms, several variants of fusion bounding, such as hot plates, ultrasonic, laser, induction, and friction stir, have been developed and matured [3].

In recent years, welding dissimilar thermoplastic materials has become a hot research topic and various techniques have been tried to acquire the robust joining. A few studies reported the fusion bounding for dissimilar thermoplastic materials with compatible molecular structures, including polyethylene (PE)-polypropylene (PP) [4], PA6-PA66 [5], polylactic acid-polyformaldehyde (PLA-POM) [6], PLA-poly (methyl methacrylate) (PMMA) [7], and PMMA-acrylonitrile butadiene styrene (ABS) [8]. These studies addressed proper welding strategies, such as welding time to ensure joint strength. So far, there are few reports on fusion bounding for incompatible polymers [9]. Fu et al. [9] successfully welded PP and PA6 by an interlayer solder sheet (ISS) prepared by blending of maleic anhydride-functionalized polypropylene (PP-g-MAH) and polyamide 6 (PA6) by an ultrasonic welding device. However, such thermoplastic systems must be processed at high processing temperatures, for example, PA-6 above 160 °C, PP in the temperature range of 230–270 °C [10]. Hence, their applications for manufacturing continuous fiber reinforced thermoplastic are limited in size and thickness due to their high melt viscosity [10].

One stablished way is a shift from melt processing to a reactive processing by using mono or oligomeric precursors, such as PA12 polymerized from *w*-laurolactam (Tm = 154 °C) using an initiator. The reactive processing of continuous fiber reinforced thermoplastic composite is performed at the temperature range of 90–250 °C. Hence, manufacturing of large composite parts, such as wind turbine blades, becomes very difficult [10]. Arkema developed an infusible thermoplastic based acrylic resin to manufacture composite parts at the temperature range of 25–90 °C depending on the variants of the acrylic resin [11]. This thermoplastic resin, known by its commercial name Elium®, possesses comparable in-plane mechanical properties to the high-performance epoxy resin after curing at room temperature [12]. Elium® matrix fiber reinforced composites can be manufactured by using the Liquid Composite Molding family such as Resin transfer Molding and vacuum infusion processes thanks to their low viscosity range from 50 mPas to 100 mPas [11,12] at room temperatures. The components of Elium® acrylic resin are 2-Propenoic acid, 2-methyl-, methyl ester or methylmethacrylate monomer (MMA), and acrylic copolymer. A radical initiator, such as a peroxide, is mixed with Elium® acrylic resin at a particular weight ratio to convert MMA to its polymer PMMA under the in situ polymerization.

In recent years, the research on the fusion bounding of Elium® composites, particularly ultrasonic welding, IR welding, and induction welding, has been conducted [13–15]. The temperature range at which it can be welded and the heating time are crucial parameters in the potential thermal degradation of the Elium® composites during welding [16,17]. To enable the use of welding for the joining, this study for first time presents hybridization of Elium® composites and a thin layer PA12 film by a reactive welding technique.

The Significace of the Study

PMMA has a remarkable transparency, but due to a low glass transition temperature around (Tg) 110 °C and high sensitivity to solvents, its applications are limited. Particularly in fusion bounding process where the temperature of exposed surfaces reaches as high as 200 °C, there is the possibility of thermal degradation of PMMA. In general, amorph polymers or semi-crystalline ones should not be welded at a temperature above 75% of their glass transition point T_g or of their melting point T_m, correspondingly [1]. There have been a few suggestions for the improvement of thermal stability of PMMA: 1- crosslinker agent: the weldability of Elium® composites with crosslinker agents are reported elsewhere by the authors [18], 2: an interposed sheet of semi-crystalline or amorphous thermoplastic polymer. In the later process, the interposed polymer sheet acts like an adhesive. The interposed polymer sheet is stacked between two similar thermoplastic composites prior the consolidation process. During consolidation, the sandwich-like polymer layers are melted together and subsequently are bounded by intermolecular diffusion [1].

The incorporation of PA-12 in the form of an interposed polymer sheet between the PMMA fiber composite plates could make it possible to overcome the thermal degradation of PMMA at high temperatures while retaining the transparency of PMMA. Polyamide-12 (PA-12) not only has a high temperature resistance with the polymer melting point of 175 °C due to the presence of strong hydrogen bounds between strings, but also has a transmission, as well as a refraction index (around 1.5) equivalent to the PMMA. It should be noted that a desired through thickness interpenetration is highly influenced by the magnitude of the viscosity difference in the interposed polymer sheet and the thermoplastic matrix resin. In addition, for a successful bounding, the interposed polymer sheet should have compatible molecular structure with the polymer matrix [19]. The addition of a third component, namely a compatibilizing agent, is a common method to enhance the compatibility between dissimilar polymers.

Compatibilization is commonly used to obtain a stable polymer mixture. A compatibilizing agent (block copolymer or graft) is introduced or formed in situ at the interface from polymers bearing functions mutually reactive chemicals. The latter process is called reactive compatibilization. The compatibilizing agent located at the interface allows a good adhesion between the dissimilar polymers thanks to the entanglements of the components of the copolymer at the interface with the polymers. It also lowers interfacial tension and prevents coalescence. Glycidyl methacrylate monomer (GMA) has been successfully used as an efficient reactive compatibilizer for reactive polymer blending, such as poly (butylene terephthalate (PBT)/polypropylene (PP) [20], PP/polycarbonate (PC) [21], PP/Polyamide 6 [22]. The presence of an epoxy group in the GMA provides reactive sites, which enables copolymerizing with a variety of monomers for modifications such as crosslinking and interfacial adhesion. All compatibilizers, which are well-known for reactive blending, are not successfully implemented to the functionalization of a liquid resin thermoplastic due to the viscosity limitation for liquid composite molding, which must be below 0.5 Pa.s. In addition, the compatibilizers should not modify the fast-curing kinetics of the acrylic-based resin such as Elium®, which is well adopted for the liquid composite molding (LCM) process. Finding an effective compatibilizer for the functionalization of a liquid resin thermoplastic that can react at their welding interface is not a trivial task. However, no reactive welding has been reported for the above systems. To fully exploit the Elium® thermoplastic composite design possibility, this study demonstrates the role of in situ functional Elium®-GMA copolymer composites.

This study proposed the functionalization of acrylics-based resin by GMA for the liquid composite molding (LCM) of functionalized continuous fibers reinforced composites and the reactive welding. For this purpose, we manufacture composite plates by a typical vacuum infusion and PA12 sheets by a thermostamping technique. Subsequently, hot-plate welding is used for the reactive welding of the interposed PA12 sheet and Elium®-GMA-Glass composite. The variable involved in this study is temperature. The experiment steps are shown in Figure 1.

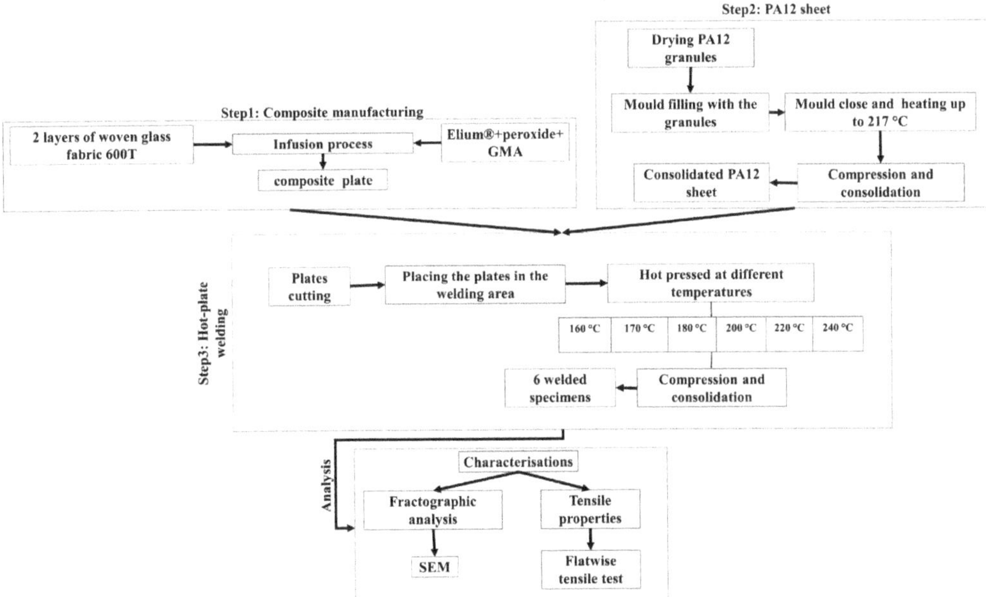

Figure 1. The experiment steps for hot-plate welding of reactively compatibilized acrylic-based composites/polyamide (PA)-12 material.

2. Materials and Methods

2.1. Fabrication Procedure

The welding was conducted at three separate steps. Step 1 was manufacturing glass fiber composite plates with an Elium® matrix. Step 2 was manufacturing PA12 sheets. Step 3 was hot-tool welding.

2.1.1. Composite Manufacturing

The Glass 600T, which is a continuous twill-weave glass fabric with an areal density of 600 g/m^2 from CHOMARAT (Le Cheylard, France), was used. Two-layered preform was formed with 350 mm × 350 mm fabric plies, leading to a thickness around one mm after vacuum bagging. The atmospheric pressure pulled the mixture of the Elium® 188XO resin with the initiator in the mass ratio of 100:2 and 5 wt% GMA (ref.779342 from Mercks, Branchburg, New Jersey, United States) from its reservoir into the preform by vacuum.

2.1.2. Thermoplastic Polymer Sheets (PA12)

A compression molding machine (Figure 2 left) was used to melt and compress the pre-weighed PA12 granules from Arkema. The sheets were fabricated with the size 100 mm × 100 mm × 1.5 mm. The fabrication steps (Figure 2 right) started with drying PA12 granules at 80 °C for 12 h. Before filling the mold with 12 g dried PA12, the lower and the upper mold were covered by polyimide RCBS 5, which is a release film for thermoforming of thermoplastic materials up to 405 °C (see Figure 2). The mold was filled with PA12 granules and was subsequently closed. The process was followed with the mold heating to 217 °C to obtain the tool temperature of 200 °C. At 200 °C, a compression pressure of 0.5 MPa was applied and held for 10 min to achieve the desired sheet thickness. Finally, the consolidated PA12 sheet was demolded.

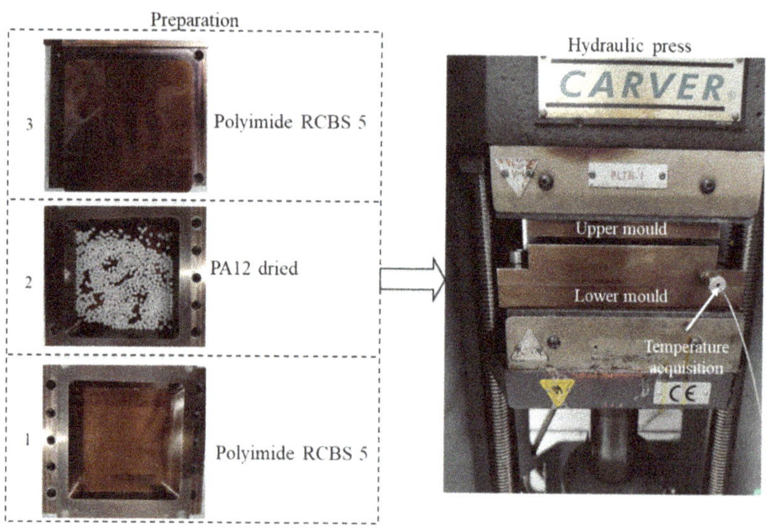

Figure 2. Preparation of PA12 sheet. 1- the lower mold covered by polyimide RCBS 5, 2- 12 g dried PA12, 3- upper mold covered by polyimide RCBS 5.

2.1.3. Assembly of PA12 Sheets with Elium® Composite Plates

For the assembly, the two 100-mm squared plates were cut from the manufactured Elium® composite, and subsequently were masked with polyimide tape on the rough side (side faced with the vacuum bag) as shown in Figure 3. The masked plates were later dried at 80 °C for 12 h. Subsequently, the plates were cleaned with Diestone and dried again at 80 °C for 30 min.

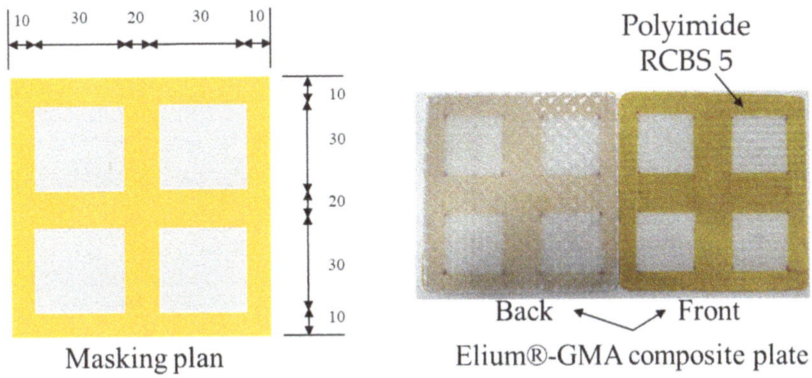

Figure 3. Masked Elium®-GMA glass fiber composite (EMG) film preparation. All dimensions are in mm.

The assembly was started with stacking the first layer of polyimide RCBS5, a masked Elium®-GMA plate, a PA12 sheet, a masked Elium®-GMA plate, and the last layer of polyimide RCBS5 (Figure 4).

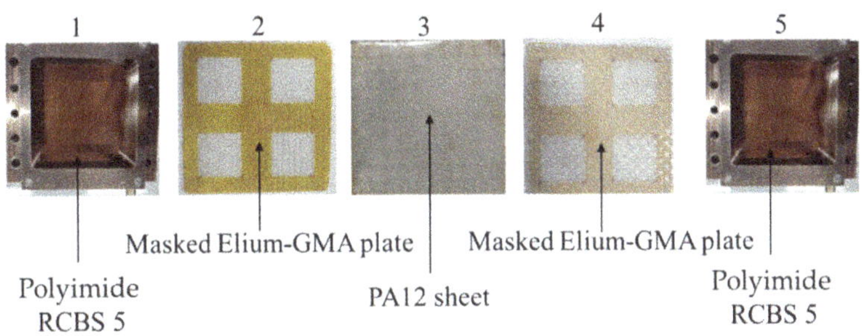

Figure 4. Sequence of layers arranged in the mold. 1- layer of polyimide RCBS5, 2- a masked Elium®-GMA plate, 3- a PA12 sheet, 4- a masked Elium®-GMA plate, 5- layer of polyimide RCBS5.

The entire assembly was hot pressed at a fixed compression pressure (0.5 MPa) and a fixed holding time (10 min). The temperature of hot-plate surfaces can be adjusted and was the only variable parameter to study the weldability of reactively compatibilized acrylic-based composites/polyamide (PA)-12 material (see Table 1).

Table 1. Temperature changes during the hot-plate welding of reactive PA12-EMG at constant hold time and constant compaction pressure. The acronym of EMG stands for Elium®-GMA-Glass composite. The samples are compressed with a compaction pressure of 0.5 MPa and held for a period of 10 min.

Sample Name	Heating Temperature (°C)
PA12-EMG 20	160
PA12-EMG 19	170
PA12-EMG 15	180
PA12-EMG 16	200
PA12-EMG 17	220
PA12-EMG 18	240

2.2. Flatwise Tensile Test

Three specimens from each welded Elium® composite laminate-GMA-PA12 were cut. The cutting configuration was shown in Figure 5A. The surface area of a specimen was

50 mm × 50 mm with an average thickness of 2.9 mm. For each of the welding scenarios (see Table 1), three iterations (Figure 5B 1–3) for the tensile test were carried out at repeatable conditions (at a temperature of 23 °C and a relative humidity of 50%).

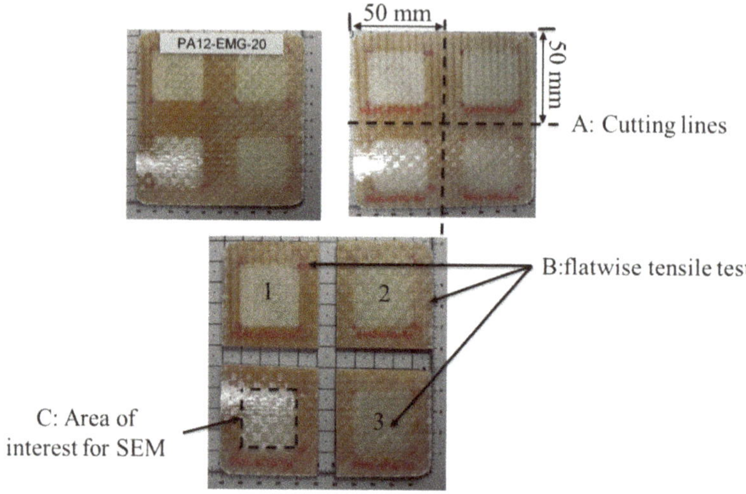

Figure 5. (**A**) Cutting configuration, (**B**) specimens for flatwise tensile test (1–3), (**C**) area of interest for Scanning Electron Microscopy (SEM).

The flatwise tensile test was conducted on the specimen according to ASTM C297 [23]. The square specimen was loaded into the universal testing machine. Figure 6A shows the whole assembly setup for the experiment, while the assembled square sample configuration under stress through the thickness is given in Figure 6B,C. The load cell of the universal Instron machine for the tensile test was adjusted to 1 kN at the displacement rate of 2 mm/min. The data were recorded in terms of the maximum load and the load-displacement curve.

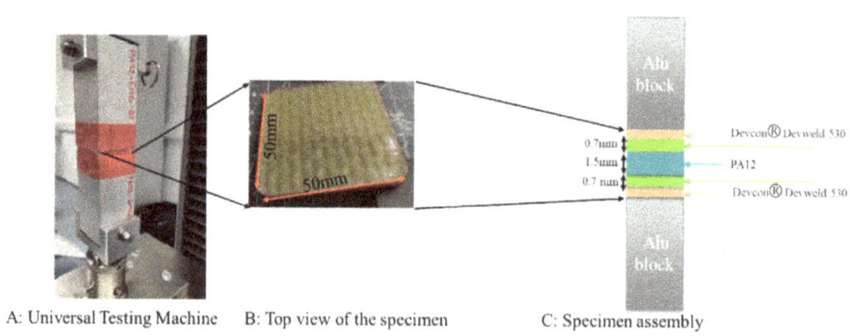

Figure 6. Schematic of assembled samples for flatwise test according to ASTM C297.

2.3. Fractographic Analysis

When the specimens are subjected to flatwise tensile test, fracture is cohesive, adhesive, or interfacial depending on the location of debonding [24]. To recognize the type of fracture, a fractographic analysis was carried out on the coupon surfaces using Scanning Electron Microscopy (SEM) (Figure 5C). SEM was a pressure-controlled FEI Quanta 200 FEG (Field Electron and Ion Company, FEI, Hillsboro, Oregon, United States). After mounting the samples in resin, they were polished with a diamond to achieve a sub-micron finish. Due to poor electrical conductivity, the samples were coated in a thin layer (about 10 nm) of gold.

3. Results and Discussion

3.1. Tensile Properties

As seen in Figure 7, at a constant hold time and constant pressure, the max load value increased with increasing the welding temperature from 160 °C to 200 °C, after which the load values tend to fluctuate between 17 MPa and 20 MPa with a further increase in the temperature. Therefore, the compatibilizer efficiency increases rapidly with increasing temperature. At 160 °C, which is below the melting point of PA12, the chain mobility was restricted, which, in turn, cannot induce the interaction of chain molecules [25]. At this temperature (160 °C), the joint of the interposed PA12 sheet and the EMG composite was broken at the little strength of 0.06 MPa due to insufficient reaction [26]. However, at 200 °C, as PA12 melting occurs [27], the welding strength of PMMA-PA12 by GMA reached 18 MPa, which is 300 times higher than the welding strength of PMMA-PA12 by GMA at 160 °C. This shows that fusion bounding between PMMA and PA12 by interfacial compatibilization was successfully achieved under hot-plate welding at 200 °C.

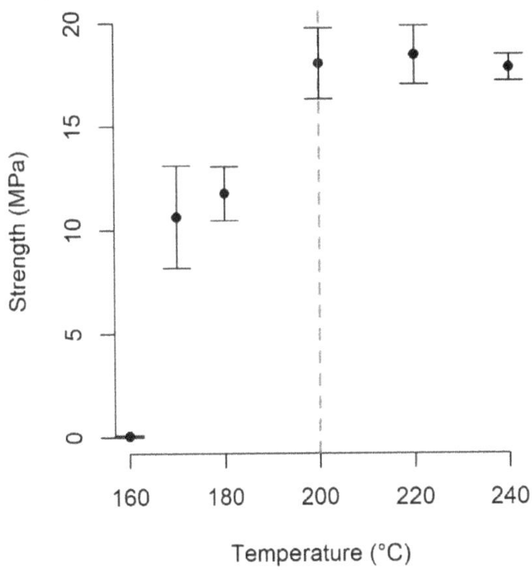

Figure 7. Variation of tensile strength versus heating temperature at constant holding time and pressure for PA12-EMG. EMG stands for Elium®-GMA-Glass composite.

3.2. Fractography

The SEM images of the flatwise tensile fractured surfaces of specimens are shown in Figures 8 and 9. The fractographic analyses revealed remarkable differences on the fracture profiles between the specimens welded in the temperature range of 160 °C to 180 °C (Figure 8) and those of 200 °C to 240 °C (Figure 9).

For the case of hot-tool reactive welding at 160 °C, Figure 8-first row shows that the surface of the adhesive is smooth and debonding occurred between the interposed PA12 sheet (labeled with number 2 in Figure 8) and the adherent (composite which is labeled with number 1). Therefore, the major fracture type is adhesive or interfacial [28]. This poor bonding quality is caused by the insufficient welding temperature. The temperature is below the melting point of PA12, and thus cannot induce the mobility of chain molecules to react with the compatibilizing agent [29].

On the other hand, the surface of the specimens welded at 170 °C in Figure 8-second row was rather rough as compared to the case of 160 °C. This is obviously due to the chain mobility of PA12. As the temperature increased to 180 °C (Figure 8-third row), the surface become rougher. The higher temperature indicates that there is the interaction of molecules at interface thanks to further mobility in the PA12 chain. It should be noted that the failure mode for the reactive welding in the temperature range of 160 °C to 180 °C was an adhesive fracture. This is well correlated to an insufficient temperature for the perfect bounding of the joints.

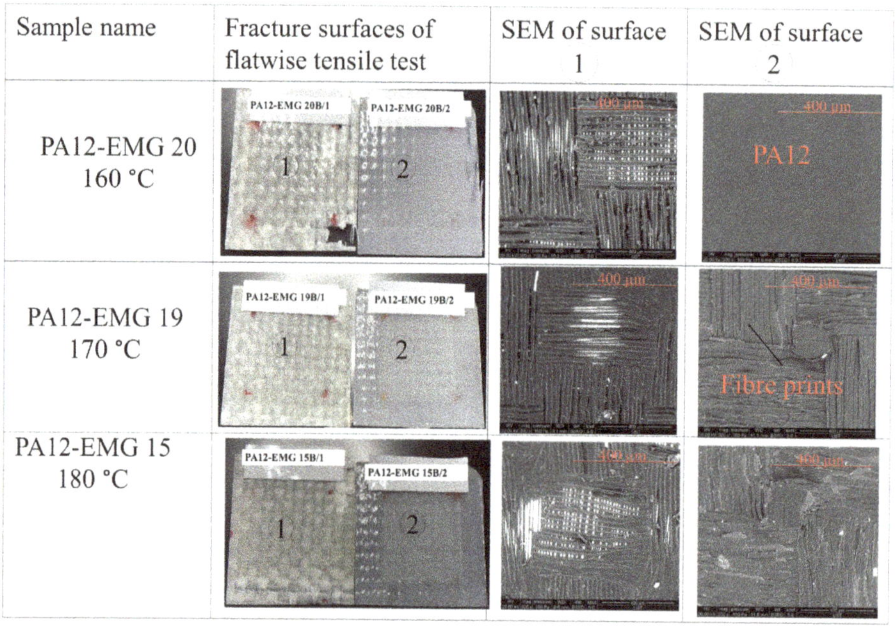

Figure 8. Fracture surfaces and SEM images of PA12-EMG specimens welded at temperatures of 160 °C, 170 °C, and 180 °C. The surface 1 stands for the EMG composite and the surface 2 stands for the interposed PA12 sheet.

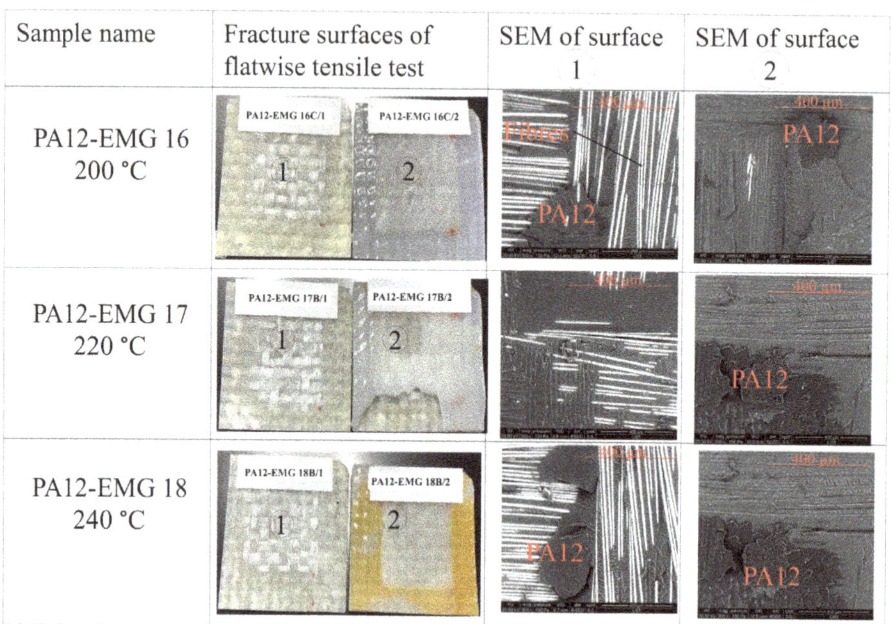

Figure 9. Fracture surfaces and SEM images of PA12-EMG specimens welded at temperatures of 200 °C, 220 °C, and 240 °C. The surface 1 stands for the EMG composite and the surface 2 stand for the interposed PA12 sheet.

The perfect bounding joints are observed in Figure 9, for the welding cases at 200 °C, 220 °C, and 240 °C. The major fracture type for the cases welds at 200 °C (Figure 9-first row) or above (Figure 9-second and third rows) is a cohesive fracture. The fractured layer (PA12, marked on the SEM images) remained on the surface of both adherend (Elium® composite plates) and the interposed PA12 sheet after debonding. Compared to the previous cases, the temperature for the cases in Figure 9 is higher, indicating the strong interaction at interfaces. Once the temperature increases to the melting temperature, the polymer flows under the compression force across the interface. Hence, the flowing polymer can react with GMA, and thus the composites and the interposed PA12 sheet can be bounded completely at 200 °C, thereby resulting in a weld [3]. However, there is no remarkable change on the surface of the fractured specimens with increasing temperature from 200 °C to 240 °C. Therefore, the desired type of fracture, which is cohesive [13], was achieved at a welding temperature of 200 °C and further increase will not change the bounding quality. This observation is supported with the results of flatwise tensile tests which were reported in the previous section.

4. Conclusions

The weldability of Elium® composite copolymerized with a glycidyl methacrylate (GMA) and PA12 as an interposed sheet were investigated by hot-plate welding. PMMA and PA12 are incompatible for welding. The results of our study on the possibility of the fusion bounding for incompatible polymers by hot-plate welding showed the following:

(1) Despite the lack of knowledge on the selection of appropriate compatibilizing agent, this study proved the effectiveness of GMA on the welding of PMMA/PA12.
(2) The highest weld strengths between the incompatible polymers PA12 and Elium® (PMMA) composite are achieved at a temperature above the melting temperature of PA12 and especially at 200 °C.
(3) The PMMA-GMA-PA12 copolymer has a key role in the enhancement of the interfacial adhesion between PMMA and PA12.

In future studies, the fusion bounding model for PMMA-PA12 interfaces will be proposed. In addition, promoting the formation of a better welding interface after consolidation will also be studied. Analysis of variance (ANOVA) is also helpful to analyze the influence of the processing pressure and hold time on the mechanical performance of resulting welded joints.

Author Contributions: H.P.: Conceptualization, Methodology, funding acquisition, Supervision, Validation, Formal analysis, Software, Visualization, Investigation, Review and Editing, M.B.: Validation, Formal analysis, Visualization, Investigation, Writing-original Draft, Review and Editing, V.B.: Formal analysis, Review and Editing, S.K.: Data Curation, Formal analysis R.V.: Data Curation, Formal analysis, Review and Editing. All authors have read and agreed to the published version of the manuscript.

Funding: This research was funded by Luxembourg National Research Fund (FNR), for funding Structural composite material for 3d Printing, SAMIA-3D under the research grant BRIDGES18/MS/13321465.

Institutional Review Board Statement: Not applicable.

Informed Consent Statement: Not applicable.

Data Availability Statement: Not applicable.

Acknowledgments: The authors greatly appreciate the support of Loic Borghini for the manufacturing of composite plates, welding, and mechanical tests from Luxembourg Institute of science and technology, MRT department for IR welding specimens.

Conflicts of Interest: The authors declare no conflict of interest.

References

1. Reis, J.P.; de Moura, M.; Samborski, S. Thermoplastic composites and their promising applications in joining and repair composites structures: A review. *Materials* **2020**, *13*, 5832. [CrossRef] [PubMed]
2. Lionetto, F.; Morillas, M.N.; Pappadà, S.; Buccoliero, G.; Villegas, I.F.; Maffezzoli, A. Hybrid welding of carbon-fiber reinforced epoxy based composites. *Compos. Part A Appl. Sci. Manuf.* **2018**, *104*, 32–40. [CrossRef]
3. Yousefpour, A.; Hojjati, M.; Immarigeon, J.P. Fusion bonding/welding of thermoplastic composites. *J. Thermoplast. Compos. Mater.* **2004**, *17*, 303–341. [CrossRef]
4. Hajideh, M.R.; Farahani, M.; Alavi, S.A.D.; Ramezani, N.M. Investigation on the effects of tool geometry on the microstructure and the mechanical properties of dissimilar friction stir welded polyethylene and polypropylene sheets. *J. Manuf. Process* **2017**, *26*, 269–279. [CrossRef]
5. Bates, P.J.; Dyck, C.; Osti, M. Vibration welding of nylon 6 to nylon 66. *Polym. Eng. Sci.* **2004**, *44*, 760–771. [CrossRef]
6. Zhang, G.; Qiu, J.; Sakai, E.; Zhou, Z. Interface investigation between dissimilar materials by ultrasonic thermal welding by the third phase. *Int. J. Adhes. Adhes.* **2021**, *104*, 102722. [CrossRef]
7. Zhang, G.H.; Qiu, J.H.; Shao, L.; Fu, X. Molecular Interdiffusion of Hauling Theory between Dissimilar Polymers Based on Novel USW. *Adv. Mat. Res.* **2011**, *221*, 289–294. [CrossRef]
8. Dashatan, S.H.; Azdast, T.; Ahmadi, S.R.; Bagheri, A. Friction stir spot welding of dissimilar polymethyl methacrylate and acrylonitrile butadiene styrene sheets. *Mater. Des.* **2013**, *45*, 135–141. [CrossRef]
9. Fu, X.; Wu, X.; Huang, G.; Li, W.; Kang, S.; Wang, L.; Luo, J.; Pan, Z.; Lu, W. Fusion Bonding Possibility for Incompatible Polymers by the Novel Ultrasonic Welding Technology: Effect of Interfacial Compatibilization. *ACS Omega* **2022**, *7*, 14797–14806. [CrossRef]
10. van Rijswijk, K.; Bersee, H.E.N. Reactive processing of textile fiber-reinforced thermoplastic composites-An overview. *Compos. Part A Appl. Sci. Manuf.* **2007**, *38*, 666–681. [CrossRef]
11. Bodaghi, M.; Park, C.H.; Krawczak, P. Reactive Processing of Acrylic-Based Thermoplastic Composites: A Mini-Review. *Front. Mater.* **2022**, *9*, 931338. [CrossRef]
12. Obande, W.; Mamalis, D.; Ray, D.; Yang, L.; Brádaigh, C.M. Mechanical and thermomechanical characterisation of vacuum-infused thermoplastic- and thermoset-based composites. *Mater. Des.* **2019**, *175*, 107828. [CrossRef]
13. Gohel, G.; Bhudolia, S.K.; Kantipudi, J.; Leong, K.F.; Barsotti, R.J. Ultrasonic welding of novel Carbon/Elium®with carbon/epoxy composites. *Compos. Commun.* **2020**, *22*, 100463. [CrossRef]
14. Murray, R.E.; Roadman, J.; Beach, R. Fusion joining of thermoplastic composite wind turbine blades: Lap-shear bond characterization. *Renew. Energy* **2019**, *140*, 501–512. [CrossRef]
15. Murray, R.E.; Penumadu, D.; Cousins, D.; Beach, R.; Snowberg, D.; Berry, D.; Suzuki, Y.; Stebner, A. Manufacturing and Flexural Characterization of Infusion-Reacted Thermoplastic Wind Turbine Blade Subcomponents. *Appl. Compos. Mater.* **2019**, *26*, 945–961. [CrossRef]
16. Wang, Y.; Rao, Z.; Liao, S.; Wang, F. Ultrasonic welding of fiber reinforced thermoplastic composites: Current understanding and challenges. *Compos. Part A Appl. Sci. Manuf.* **2021**, *149*, 106578. [CrossRef]
17. Unnikrishnan, T.G.; Kavan, P. A review study in ultrasonic-welding of similar and dissimilar thermoplastic polymers and its composites. *Mater. Today Proc.* **2022**, *56*, 3294–3300. [CrossRef]
18. Perrin, H.; Bodaghi, M.; Berthé, V.; Vaudenmont, R. On the effect of multifunctional methacrylate monomer content on the IR weldability of acrylic-based glass fibre composites. *Polym. Compos.* 2022; *Under review*.
19. Smiley, A.J.; Halbritter, A.; Cogswell, F.N.; Meakin, P.J. Dual polymer bonding of thermoplastic composite structures. *Polym. Eng. Sci.* **1991**, *31*, 526–532. [CrossRef]
20. Tsai, C.-H.; Chang, F.-C. Polymer blends of PBT and PP compatibilized by ethylene-co-glycidyl methacrylate copolymers. *J. Appl. Polym. Sci.* **1996**, *61*, 321–332. [CrossRef]
21. Zhihui, Y.; Yajie, Z.; Xiaomin, Z.; Jinghua, Y. Effects of the compatibilizer PP-g-GMA on morphology and mechanical properties of PP/PC blends. *Polymer* **1998**, *39*, 547–551. [CrossRef]
22. Tedesco, A.; Barbosa, R.; Nachtigall, S.M.B.; Mauler, R.S. Comparative study of PP-MA and PP-GMA as compatibilizing agents on polypropylene/nylon 6 blends. *Polymer Testing.* **2002**, *21*, 11–15. [CrossRef]
23. INSTRON Composite Test Fixture ASTMC297. 2019. Available online: https://www.astm.org/standards/c297 (accessed on 1 January 2023).
24. Castellani, L.; Castiglioni, A.; Ferri, D.; Baraldi, A.; Martinelli, L.; Rotunno, S. Rate effects on adhesion energy between polyethylene films and different substrates. *Eng. Fract. Mech.* **2015**, *149*, 387–401. [CrossRef]
25. Stokes, V.K. A phenomenological study of the hot-tool welding of thermoplastics Part 3. Polyetherimide. *Polymer* **2001**, *42*, 775–792. [CrossRef]
26. Meng, X.; Huang, Y.; Cao, J.; Shen, J.; Santos, J.F. Recent progress on control strategies for inherent issues in friction stir welding. *Prog. Mater. Sci.* **2021**, *115*, 100706. [CrossRef]
27. Ageorges, C.; Ye, L. State of the Art in Fusion Bonding of Polymer Composites. In *Fusion Bonding of Polymer Composites;* Springer: London, UK, 2002; pp. 7–64. [CrossRef]

28. Katsiropoulos, C.; Chamos, A.N.; Tserpes, K.I.; Pantelakis, S.G. Fracture toughness and shear behavior of composite bonded joints based on a novel aerospace adhesive. *Compos. B Eng.* **2012**, *43*, 240–248. [CrossRef]
29. McIlroy, C.; Olmsted, P.D. Disentanglement effects on welding behaviour of polymer melts during the fused-filament-fabrication method for additive manufacturing. *Polymer* **2017**, *123*, 376–391. [CrossRef]

Disclaimer/Publisher's Note: The statements, opinions and data contained in all publications are solely those of the individual author(s) and contributor(s) and not of MDPI and/or the editor(s). MDPI and/or the editor(s) disclaim responsibility for any injury to people or property resulting from any ideas, methods, instructions or products referred to in the content.

materials

MDPI

Communication

On the Analyses of Cure Cycle Effects on Peel Strength Characteristics in Carbon High-T_g Epoxy/Plasma-Activated Carbon PEEK Composite Interfaces: A Preliminary Inquiry

Henri Perrin, Régis Vaudemont and Masoud Bodaghi *

Luxembourg Institute of Science and Technology (LIST), 5 rue Bommel, L-4940 Hautcharage, Luxembourg; henri.perrin@list.lu (H.P.); regis.vaudemont@list.lu (R.V.)
* Correspondence: masoud.bodaghi@list.lu; Tel.: +352-2758884575

Abstract: In this study, a high-T_g aerospace-grade epoxy composite plate was co-curing welded using a unidirectional PEEK thermoplastic carbon fibre tape to develop advanced composite joints. To account for the surface roughness and the weldability of carbon–epoxy/carbon–PEEK composites, plasma treatments were performed. The co-curing was conducted by the following steps: each treated thermoplastic tape was first placed in the mould, and followed by nine layers of dry-woven carbon fabrics. The mould was sealed using a vacuum bag, and a bi-component thermoset (RTM6) impregnated the preform. To understand the role of curing kinetics, post-curing, curing temperature, and dwell time on the quality of joints, five cure cycles were programmed. The strengths of the welded joints were investigated via the interlayer peeling test. Furthermore, cross-sections of welded zones were assessed using scanning electron microscopy in terms of the morphology of the PEEK/epoxy interphase after co-curing. The preliminary results showed that the cure cycle is an important controlling parameter for crack propagation. A noticeable distinction was evident between the samples cured first at 140 °C for 2 h and then at 180 °C for 2 h, and those cured initially at 150 °C for 2 h followed by 180 °C for 2 h. In other words, the samples subjected to the latter curing conditions exhibited consistently reproducible results with minimal errors compared to different samples. The reduced errors confirmed the reproducibility of these samples, indicating that the adhesion between CF/PEEK and CF/RTM6 tends to be more stable in this curing scenario.

Keywords: co-curing bonding; plasma treatment; PEEK/epoxy interphase

Citation: Perrin, H.; Vaudemont, R.; Bodaghi, M. On the Analyses of Cure Cycle Effects on Peel Strength Characteristics in Carbon High-T_g Epoxy/Plasma-Activated Carbon PEEK Composite Interfaces: A Preliminary Inquiry. *Materials* **2023**, *16*, 7340. https://doi.org/10.3390/ma16237340

Academic Editor: Mattia Biesuz

Received: 20 October 2023
Revised: 20 November 2023
Accepted: 21 November 2023
Published: 25 November 2023

Copyright: © 2023 by the authors. Licensee MDPI, Basel, Switzerland. This article is an open access article distributed under the terms and conditions of the Creative Commons Attribution (CC BY) license (https://creativecommons.org/licenses/by/4.0/).

1. Introduction

Composite materials, both thermoplastic and thermoset, have found extensive applications in land transportation, aerospace, and marine structures, gradually replacing their traditional metallic counterparts [1,2]. This shift is primarily attributed to the superior strength and stiffness-to-weight ratios offered by composite materials when compared to metals. The superior specific properties of composites, in contrast to metals, result in reduced weight, increased payload capacity, extended operational range, and improved overall mechanical performance of structures.

There has been enormous attention from researchers and the aerospace industry alike to the use of thermoplastic composites (TPCs). Advanced thermoplastic composites present numerous processing and mechanical performance benefits compared to thermoset composites. These advantages include high damage tolerance, exceptional corrosion and solvent resistance, elevated fracture toughness, superior impact resistance, commendable fatigue resistance, economical storage costs, and an indefinite shelf life [2]. An example of this is the thousands of press-formed carbon fibre/polyetheretherketone (CF/PEEK) clips that were previously built into the A350 and Boeing 787 aircraft [3].

Compared to their thermoplastic counterparts, the jointing and assembly of CF/Epoxy composites pose a significant challenge for producing large-scale complex composite structural parts. Due to the high level of cross-linking, fusion bonding is no longer viable for cured CF/epoxy composites [2,4–7]. Instead, conventional methods such as mechanical fastening and adhesive bonding are employed as alternative solutions. The CF/PEEK is now attached to the CF/epoxy fuselage skin via mechanical fastening, which is a well-established technique for aircraft metallic structures [8]. However, these methods have their own inherent drawbacks when it comes to joining and assembling composite parts. With mechanical fastening, challenges typically arise from stress concentrations, potential delamination induced by drilling, potential galvanic corrosion, limitations in lightweight design, and the time-consuming, labor-intensive nature of the process [9–11]. In the case of adhesive bonding, extensive surface preparation and long curing cycles are necessary, making it inherently incompatible with the requirements of mass production [4,11,12].

Fusion bonding is considered an optimal technique for joining thermoplastic matrix composites. It involves three key stages: first, heating the thermoplastic matrices at the interface to achieve a melted state; next, applying pressure to the paired parts to facilitate the inter-diffusion of polymer chains; and finally, allowing the composite to cool down and regain its mechanical properties [13]. A relevant question is how the existing techniques can be applied for the welding of dissimilar composites, i.e., TPC and thermoset composite (TSC) combinations.

To expand the use of fusion bonding in thermoset polymer matrix composites, it is necessary to coat the interface of the laminates with a layer of thermoplastic material [14]. However, the thermoplastic hybrid interlayer method may encounter challenges in achieving the desired through-thickness inter-penetration depth of the woven roving fiber cloth. This difference in melt viscosity can be attributed to the significant contrast between the thermoplastic film and the thermoset matrix resin. Moreover, the elevated temperature needed to melt the thermoplastic hybrid interlayer or film may result in the degradation of the thermoset matrix resin. One of the most common solutions to increase the occurrence of interfacial adhesion is using a thermoplastic interlayer that is compatible with the thermoset resin in TSC [9]. Some examples of such interlayer thermoplastic resins are polyetherimide (PEI), polysulfone (PSU), and polyethersulfone (PES), which are of an amorphous nature [14]. In addition to the compatibility of the inter-layer thermoplastic resins with the thermoset matrix, a few other factors must also be considered to select the appropriate inter-layer thermoplastic resin: mechanical and environmental performance in applications, techniques used to incorporate the interlayers, and weldability of the interlayer thermoplastic resin without thermoset degradation. To achieve a sufficient high-strength bond without influencing the thermoset cure cycle, the chemical and physical properties of the interlayer and thermoset must also be considered. Moreover, issues such as resistance to aircraft solvents and moisture must also be addressed [14]. Consequently, it typically requires intermediate pretreatment involving plasma and corona discharge to enhance its interfacial adhesion with the thermoset matrix. One significant advantage of plasma treatment is that it preserves material integrity, reducing the likelihood of fibre degradation compared to chemical treatment. Additionally, plasma treatment is a clean and dry process, mitigating environmental concerns typically associated with chemical modification [15]. The surface of the TPC can be treated by the atmospheric plasma to generate a certain surface roughness that will cause mechanical interlock between the TSC and TPC during the welding process [16]. This solution does not require the interlayer and makes direct welding of a TPC on the treated surface of TSC possible. Understandably, some of the main points of concern in this solution are the weld strength and durability of the thermoplastic and thermoplastic connection.

This study proposes a new technique to achieve a high-strength connection of the hybrid interface by co-curing the unidirectional carbon fiber-reinforced thermoplastic tape with the thermoset composite without the requirement of an interlayer thermoplastic. This enables us to directly test the interface using the peeling test without an additional welding process, like infrared welding, which could be responsible for thermal degradation of the TSC substrate.

Co-curing, a method for joining polymer composite joints, involves simultaneous curing, offering cost-effectiveness compared to co-bonding or secondary bonding techniques [17–19]. Co-bonding implies curing stacked prepregs alongside other parts, while secondary bonding involves bonding cured parts using adhesive. After establishing the concept, the surface of the CF/epoxy composite laminates underwent a coating process using various thermoplastic binders in either powder or film form. Subsequently, the composites were co-cured through hot pressing. It was anticipated that the elevated curing temperature would induce the melting of the thermoplastic binder, allowing it to blend with the epoxide resin in the CF/epoxy prepreg under vacuum and/or external pressure. This blend was then consolidated through the curing of the epoxide resin and the thermoplastic binder. As for the heating element, a carbon-fiber-reinforced binder (CF/binder) prepreg interlayer, comprising a single layer of carbon fabric sandwiched between two layers of binder film, was prepared through hot pressing [20].

Several studies have examined thermoset-to-thermoset composite joints. Moretti et al. [21] investigated process-induced strains during autoclave co-curing, co-bonding, and secondary bonding of epoxy composite laminates, noting minimal warpage in the co-curing process. Hasan et al. [22] produced full-scale wing demos joined by co-curing or secondary bonding, revealing significantly less laminate warpage in co-curing bonded joints compared to secondary bonded ones. Furthermore, co-curing bonded demos satisfied engineering tolerances without any defects or anomalies. In addition to its manufacturing quality and accuracy, several studies reported adequate or excellent mechanical properties of co-curing bonded joints [23–25]. For instance, Dhilipkumar and Rajesh [25] observed 67% and 52% higher lap shear strengths in co-curing bonded joints compared to co-bonding joints and secondary bonding joints, respectively. Kim et al. [24] reported much higher pull-off strengths in co-curing hat-stiffened panels than those manufactured by co-bonding and secondary bonding techniques. Based on these findings, co-curing emerges as a promising technique for composite joining, offering good structural integrity with minimal curing cycles.

Although there is growing understanding regarding the co-curing of thermoset-to-thermoset composite bonding, the application of this method for joining carbon fiber/PEEK tapes and aerospace carbon fiber/epoxy composites through a co-curing process without an interlayer remains limited. Quan et al. [26] offer insights into the creation of composite joints featuring robust structural integrity and thermal stability through the co-curing bonding of CF/epoxy composites and CF/PEEK tapes. Employing a UV-irradiation technique, the researchers treated the surfaces of PEEK films, a method demonstrated to notably amplify their adhesive properties with epoxies [27].

Given the prevalent use of CF/PEEK and CF/epoxy composites in aerospace applications, this paper concentrates on gaining additional insights into the co-curing process of these materials. The study explores CF/PEEK tapes as substitutes for epoxy adhesives in bonding aerospace carbon fibre/epoxy composites via co-curing. Plasma treatment was employed to modify the PEEK film surfaces, and the investigation encompassed peel testing of the composite joints bonded by CF/PEEK tapes.

2. Experimental Details

2.1. Materials

For the present investigation, a CF/PEEK tape in its amorphous form (PEEK A) was selected, as indicated in Table 1 [28–30]. The employed thermoset resin was RTM6-2, a dual-component resin designed for resin transfer moulding and infusion processes and to meet the demands of the aerospace sector (Table 2) [31,32]. The provided reinforcing fiber came from Hexcel, specifically the HexTow AS4C (Table 3), which consists of a continuous twill-woven carbon fabric derived from PAN [33].

Table 1. Nominal specification of CF/PEEK tape: T_g is glass transition temperature [28–30,34].

Commercial Name	Characteristics					
	Thickness (μm)	Fibre Fraction Volume (%)	T_g (°C)	Recrystallisation Temperature (°C)	Crystallinity (%)	Manufacturer
Aptiv®2000	25	54	143	160	8	Victrex®

Table 2. Nominal specification of RTM6-2 [31,32,35].

Commercial Name	Characteristics				
	Recommended Cure Cycle	T_g (°C)	Gel Time at 140 °C (min)	Resin Injection Temperature (°C)	Manufacturer
HexFlow® RTM6-2	120 min at 180 °C	High	95	80 under vacuum/ low pressure	Hexcel®

Table 3. Nominal specification of twill-woven carbon fabric for manufacturing of a composite laminate.

Style	Material	Number of Fibre	Weave Pattern	Areal Density
HexTow AS4C	Carbon fibre	3K	2 × 2 twill	200 g/m²

2.2. Plasma Treatment

Plasma treatments were conducted utilizing an atmospheric pressure reactor based on dielectric barrier discharge (DBD) principles. In essence, plasma was initiated within a DBD setup with a 3 mm gap between two electrodes. Each of these high-voltage aluminium plate electrodes was shielded by a glass plate measuring 3.25 mm in thickness. The plasma discharge was induced through an AC power supply configured to 450 W and 6 kHz, accompanied by a gas mixture of 80% nitrogen (N_2) and 20% oxygen (O_2). The upper electrode was moved consistently at a speed of 4 m/min over the lower electrode. The deposition duration was established at 1 min. For those interested in gaining a more comprehensive understanding of surface treatment duration, the publication [16] by the same author is recommended. The plasma treatment process necessitates strict adherence to health and safety considerations to safeguard personnel and maintain regulatory compliance. Chemical exposure poses a significant concern, as reactive gases utilised in the plasma treatment may lead to respiratory issues, requiring proper ventilation and exposure monitoring. High voltages involved in the operation demand thorough training and the implementation of safety interlocks to prevent electric shocks. Personal protective equipment, including gloves and eye protection, is imperative to shield against chemical exposure and potential splashes.

2.3. Co-Curing Process

A descriptive schematic of the co-curing process is shown in Figure 1. Liquid resin infusion (LRI) was utilized for the co-cure processing of the joint interface between CF/PEEK tape and thermoset matrix composites reinforced with woven carbon fibers (Figure 1a). The process started by laying up nine plies of twill-woven carbon fabric from Hexcel (Table 3).

The temperature of the samples was monitored with a dielectric sensor positioned at one end of the preform (Figure 1b). Following this, six CF/PEEK tapes were positioned next to each other on top of the polyimide film (Figure 1c). The CF/PEEK (Aptiv®2000) was supplied by Victrex® (Lancashire, UK). The matrix has a glass transition temperature T_g = 143 °C and a recrystallization temperature T_C = 160 °C. The tapes have an initial fibre volume fraction (V_f) close to 54 vol% [34]. After placing nine layers of twill-woven carbon fabric on top, the assembly was covered on the upper side with a flow mesh and a flexible bag. A vacuum was applied to remove air from the assembly (Figure 1d). The two-component RTM6-2 [35] was drawn into the assembly at temperatures of 80 °C from the reservoir (Figure 1e). The RTM6 resin is a single-component polyepoxide resin widely used in the aerospace industry. This resin is employed in the manufacturing processes of composites through liquid methods, notably in the LRI process or through injection on reinforcement, such as in resin transfer moulding (RTM).

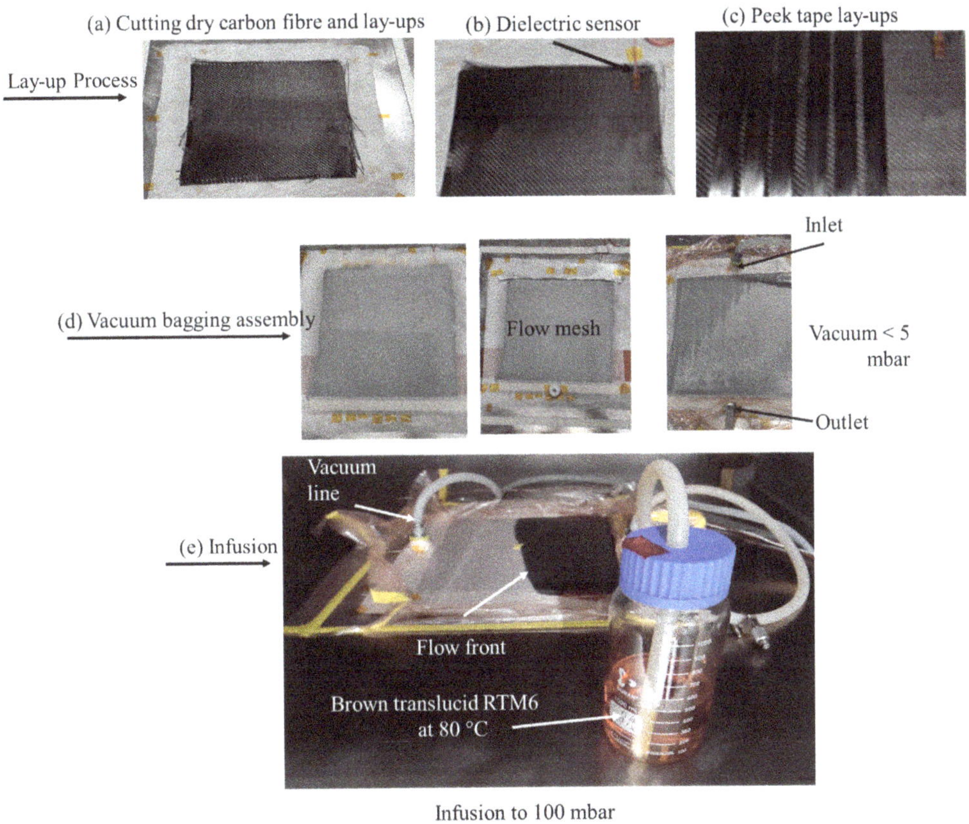

Figure 1. Manufacturing processes.

Following the impregnation process, a total of five distinct cure cycles (Figure 2) were established to investigate the influence of curing cycles on bonding strength. A ramp between 80 °C and the final set temperatures for four distinct cure cycles was 2 °C/min except for the segment 20 min + 150 °C, which was 1.2 °C/min. Then, each temperature segment remained constant for a given time (which varied between 20 min and 2 h) before a cooling ramp at 2 °C/min. For each curing cycle, six (6) samples were tested.

In view of the comparison of TTT diagrams (a time–temperature–transformation diagram obtained) with three sets of parameters from the literature [36,37], we can first observe the similarity in the appearance of the obtained diagrams. However, significant differences exist, especially in terms of time, between the obtained TTT diagrams. This reflects the reality of dispersion in reaction rates as well as the variability in the determination of kinetic parameters. Nevertheless, these graphs provide insight into the behaviour of the resin and assist in predicting the progress achieved by a curing cycle. These graphs highlight the role of vitrification concerning the advancement and reaction kinetics of the resin, with the kinetics being markedly different on either side of vitrification. Most of the literature focused on investigating the reactive characteristics of pure RTM6 has limited practical applicability, as it primarily addresses a commercial system that is never utilized without the reactive binder—a crucial component for achieving the desired properties [38]. Consequently, optimising manufacturing cure cycles was highly valuable. The manufacturing process can be divided into two phases. First, the resin is injected (or drawn) through a dry fabric, known as the impregnation phase. In the second phase, the resin is cured by applying a temperature cycle. Understanding the curing kinetics of a resin is crucial for adjusting the composite curing cycles, not only to match the specific requirements of the part being produced but also to obtain good adhesion between the thermoset and thermoplastic counterparts. Five curing cycles (Figure 2) as a function of temperature and time were, therefore, chosen based on the vitrification curve [39].

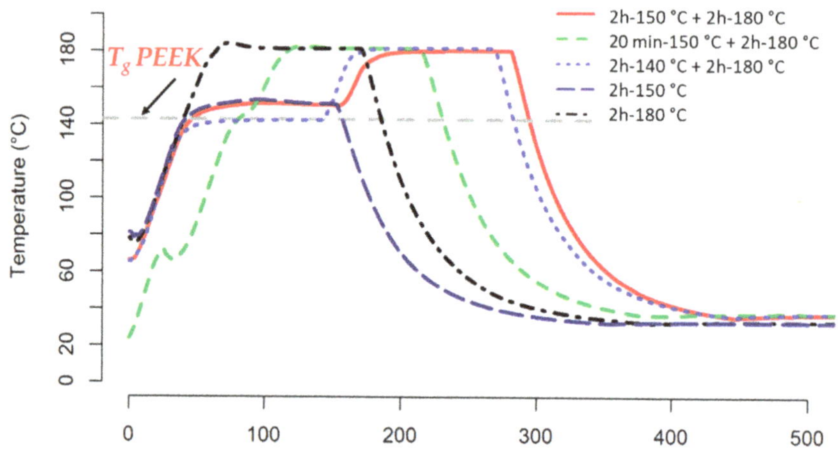

Figure 2. Five different curing cycles.

2.4. Peel Tests

The 90° peel adhesion test according to ILNAS-EN 28510-1:2014 [40] serves the purpose of evaluating the force required to separate the two components that were joined through co-curing. The outcome of this test, akin to bond strength, is denoted as N (force to de-bond)/25.4 mm (tape width). The test was conducted at a speed of 0.84 mm/s and repeated six times. The freely moving table was linked to the crosshead via a pulley and rope mechanism. This arrangement ensured that the table's lateral movement matches the crosshead's motion, thereby maintaining a consistent 90° angle between the two components. It is crucial to emphasise that the untreated PEEK/C tape configuration exhibits inadequate adhesion post-demoulding, making peeling tests unviable.

After the peel test, the fracture surfaces were examined using scanning electron microscopy (SEM). The SEM analysis was performed with an FEI Quanta FEG 200 scanning electron microscope from the FEI Company (Hillsboro, OR, USA), operating under pressure-controlled conditions.

3. Results and Discussion

3.1. Peel Tests

Figure 3a–e presents the relationship between peel force and displacement for separately co-cured samples. In general, each curve displays three distinct phases:

1. Initially, there's an increase in peeling force until the interface starts to propagate.
2. Once the interface begins to propagate, there's a slight drop in peeling force. As cracks propagate between adhered materials, the effective load transfer diminishes, resulting in a reduction in peeling force. It is important to note that Xu et al. (1992) [41] characterized a viscoelastic plate resembling a cantilever beam subjected solely to bending. Their findings indicate that interfacial toughness initially rises and subsequently declines with an increase in crack propagation velocity. This implies the potential presence of peak viscoelastic energy dissipation occurring at an intermediate crack velocity [42].

3. This is followed by a peeling process where the peeling force stabilises, represented by the average force during peeling (indicated by dashed red lines in Figure 3) in our experiments.

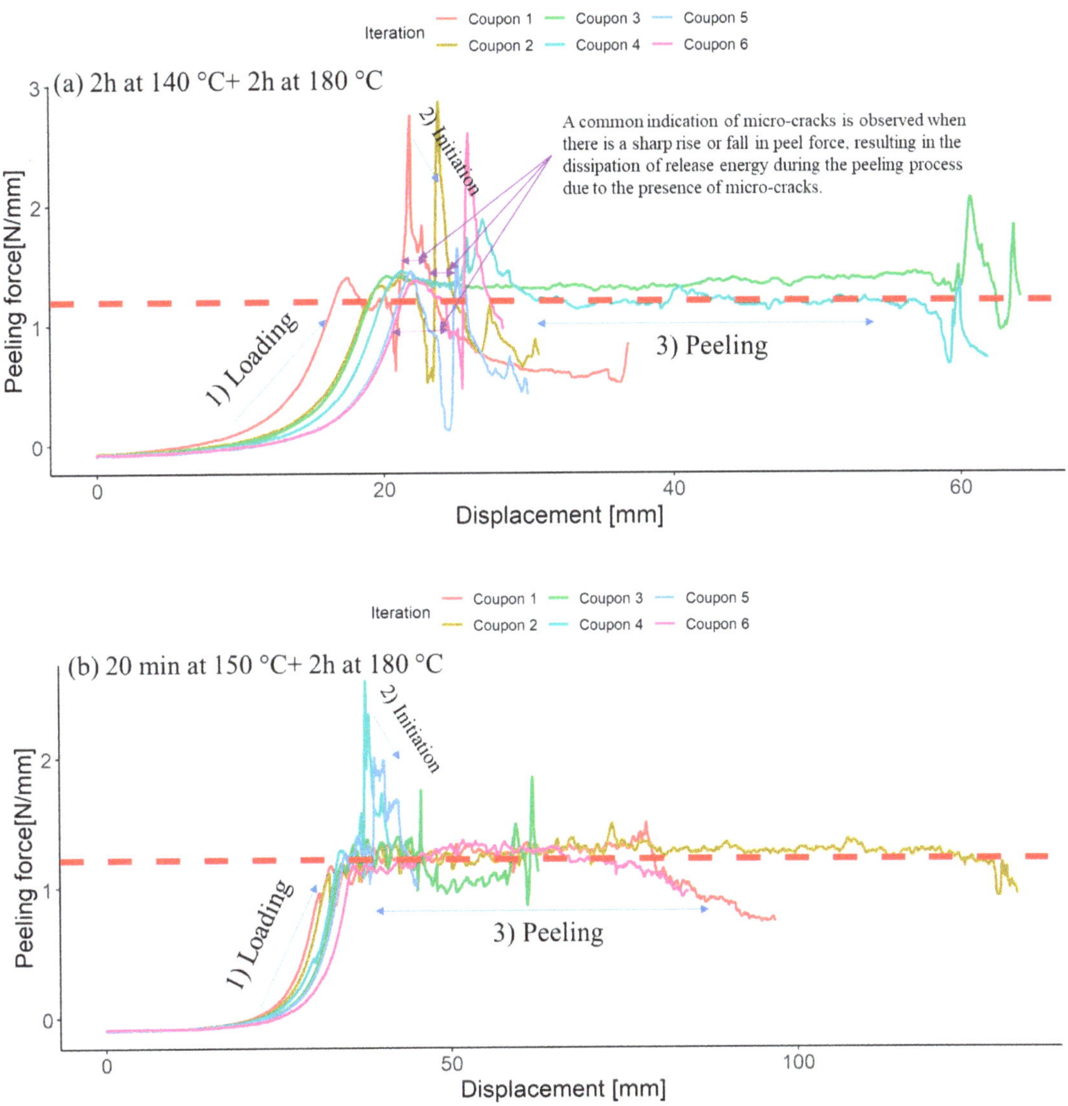

Figure 3. *Cont.*

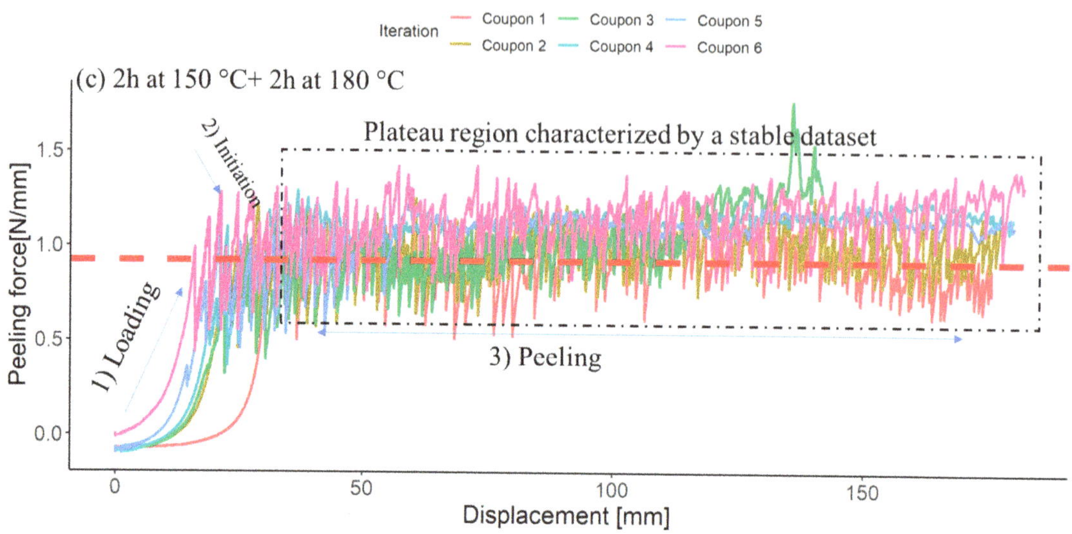

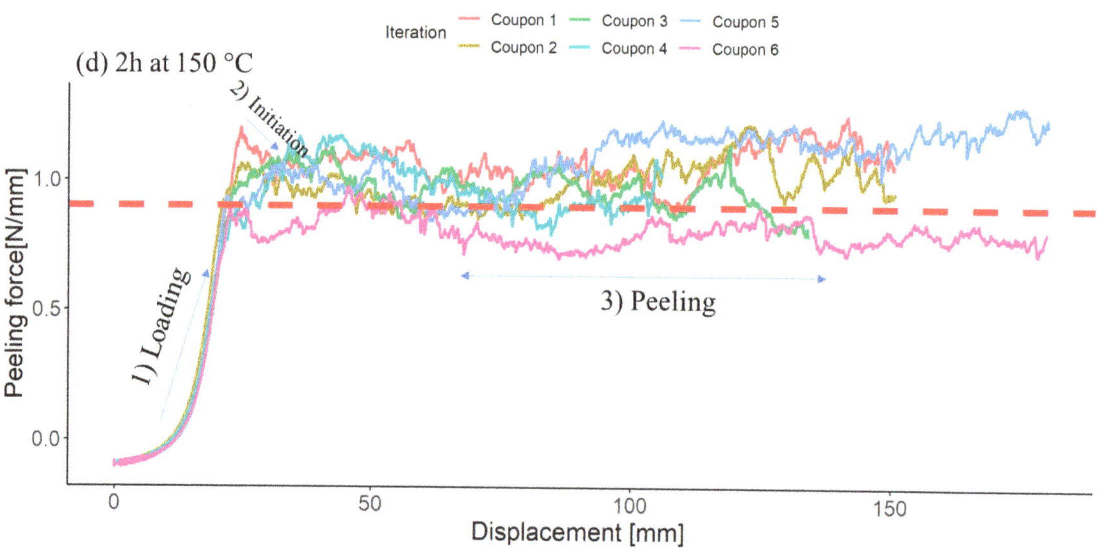

Figure 3. *Cont.*

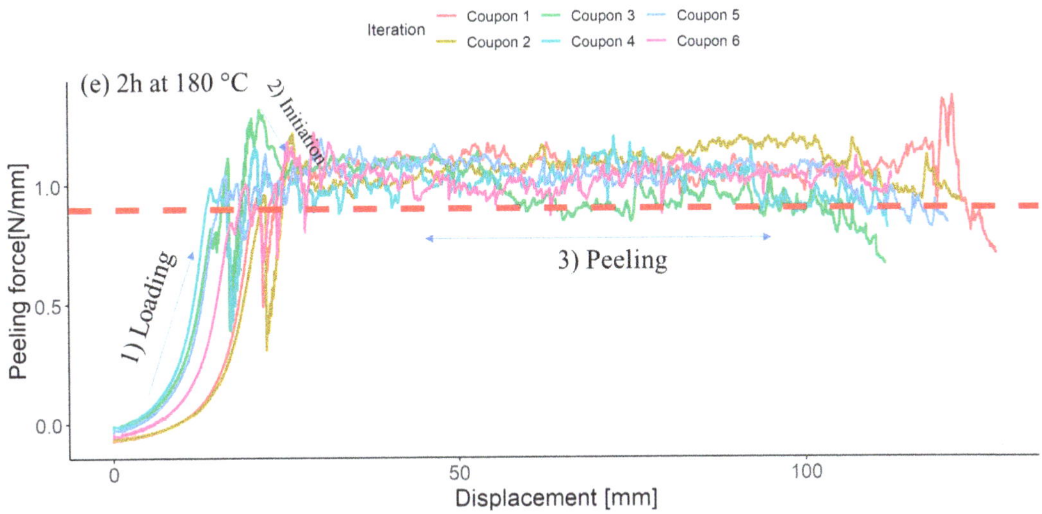

Figure 3. Load displacement for the samples cured at different dwell temperatures and time (six tests per curing cycle): (**a**) 2 h at 140 °C + 2 h at 180 °C, (**b**) 20 min at 150 °C + 2 h at 180 °C, (**c**) 2 h at 150 °C + 2 h at 180 °C, (**d**) 2 h at 150 °C, (**e**) 2 h at 180 °C. The red dashed line represents the computed average peeling force derived from multiple iterations.

Since the three distinct phases were not easily recognisable, the boundaries of these three phases are indicated by blue arrows for each curing cycle scenario in Figure 3. The force–displacement curves acquired from these tests exhibit diverse patterns contingent upon the cure cycles. Peel arms characterised by heterogeneity generate periodic peaks in the peel curve, exemplified by the peak force, denoted as Fmax in Figure 3a,b. Distinct curve shapes can be attributed to the type or structure of the heterogeneity. In contrast, debond propagation in samples cured at 150 °C for 2 h and at 180 °C for 2 h often displays temporally unstable, or stick–slip crack behaviour—sometimes referred to as "shocky" or "zippy" behaviour, as depicted in Figure 3c. This intricate phenomenon is influenced by multiple factors, including constitutive material properties, test rate, test temperature, specimen and load-train compliance, and system inertia [43]. The associated peel force versus displacement curve in Figure 3c illustrates the intermittent stick–slip growth of the crack. The reasons behind the temporal instability of crack propagation typically involve intricate, multi-physics interactions [44,45], details of which are beyond the scope of this paper.

It is evident from Figure 3 that each curing cycle resulted in a significantly different curve for the peel force versus displacement. The notable difference is apparent between the samples cured initially at 140 °C for 2 h and then at 180 °C for 2 h, and the samples cured initially at 150 °C for 2 h and 180 °C for 2 h. On the other hand, the samples cured initially at 150 °C for 2 h and 180 °C for 2 h showed reproducible results, with the least number of errors among the different samples. Fewer errors confirmed that the samples are reproducible, and the adhesion between CF/PEEK and CF/RTM6 tends to be more stable. The CF/PEEK-RTM6 samples co-cured first at 140 °C for 2 h and then at 180 °C for 2 h (Figure 3a) exhibited significant variability in results without reaching a stable state. This suggests a lack of valid adhesion between PEEK and epoxy.

In the case of co-curing initially at 140 °C for 2 h and subsequently at 180 °C for 2 h, a typical peel result revealed significant force oscillations due to staged crack development, as illustrated in Figure 3a. Figure 3a clearly depicts a notable reduction in peel force [46]. A comparable trend, albeit more moderate, was observed for the co-curing scenario, where the initial step involved curing at 150 °C for 20 min, followed by a subsequent curing at 180 °C for 2 h. This observation emphasises how curing conditions, like temperature and time, impact interphase formation and peel strength. Additionally, the stable crack propagation typically seen in unidirectional (UD) reinforced materials during testing isn't observed here, as we are dealing with woven fabric composites, particularly tough thermoplastic composites like CF/PEEK [47]. Unlike the previous observation, the results for samples cured at 150 °C for 2 h and 180 °C for 2 h showed a stable dataset. The oscillated crack propagation, especially in samples cured at 150 °C for 2 h and 180 °C for 2 h, is related to collective interactions of contact areas. In this process, when peeling occurs, a crack initiates and moves through the interface faster than the peel arms can move. When the peel arms loosen, the crack stops, and this cycle repeats, creating the characteristic stick–slip pattern [48] seen in Figure 3c. This unstable stick–slip crack growth is associated with structural transitions, such as moving from cohesive to interfacial failure or between different interfacial failure modes, confirming the formation of the PEEK and RTM6 interfaces [49]. Notably, the details of the stick–slip pattern depend on the dwell temperature relative to the glass transition temperature (T_g) of PEEK. Above the T_g, epoxy polymeric chains/PEEK are in a mobile state and can further dissolve/diffuse during the co-curing process, resulting in a thicker interface.

Figure 3d,e also demonstrate that a single-dwell curing cycle may halt the diffusion process, leading to weaker stick–slip behaviour and lower displacement compared to two dwell curing cycles. This effect becomes more pronounced with longer dwell times (2 h) and higher temperatures (150 °C). Specifically, a 2 h dwell at 150 °C and 2 h at 180 °C allow for longer mobility and diffusion times after reaching the nominal gel point. However, when considering the 20 min dwell time, these graphs emphasise the distinct impact of vitrification on the resin's reaction kinetics. It is evident that the kinetics vary significantly on either side of the vitrification point, depending on the duration. Specifically, at 150 °C, vitrification begins after approximately 100 min [36].

The results of the maximum mean peel force were compared statistically with respect to the curing cycle, as shown in Table 4. The curing cycle clearly exhibited a significant impact. While the peel force for the curing cycle of 2 h at 140 °C + 2 h at 180 °C was the highest compared to other curing cycles, it also showed the highest coefficient of variation, at 22%, confirming a high scatter and oscillation in the peel test results. On the other hand, although the mean peel forces for the samples cured for 2 h at 150 °C + 2 h at 180 °C were more consistent, they still showed a high coefficient of variation compared to the single-dwell curing cycle. This is primarily because one out of six samples showed a sudden increase in peel force due to an earlier failure.

Table 4. Statistical analysis of maximum peel force of six iterations for each corresponding curing cycle.

Curing Cycle	Mean of Maximum Peel Forces (M) (N/mm)	Standard Deviation (sd)	Coefficient of Variation (sd/M × 100)
2 h at 140 °C + 2 h at 180 °C	2.10	0.5	21.9
20 min at 150 °C + 2 h at 180 °C	1.71	0.45	25.4
2 h at 150 °C + 2 h at 180 °C	1.37	0.2	15.2
2 h at 150 °C	1.17	0.11	9.7
2 h at 180 °C	1.22	0.071	5.7

3.2. *Microscopic Investigation of the Interfaces of the Joints*

To further explore the microscopic assessment of cohesive strength at the interface, Figure 4 showcases a spectrum of failure modes, ranging from adhesive and cohesive failures within the co-cured interface to the potential failure of the thermoplastic tape. The observed failure mode varies depending on the specific dwell cure cycles employed. For the single dwell cure cycle (Figure 4a, 2 h at 180 °C), we identified a partial cohesive failure through the thermoplastic composite (TPC), distinguished by a notably rough surface. Additionally, as seen in Figure 4a, adhesive failure was observed, characterized by a smooth surface. One explanation for the adhesive failure is the highly crosslinked epoxy, which obstructed the formation of a complete interface. In the case of two dwell cure cycles, it is apparent that the surface was rougher compared to that of a single dwell cure cycle (Figure 4b,c). Nevertheless, adhesive failure on the TPC side was noticeable when curing at 140 °C for 2 h followed by curing at 180 °C for 2 h. Conversely, for the scenario involving curing at 150 °C for 20 min and then 2 h at 180 °C, the dominant mechanism was cohesive failure through TPC. Nonetheless, as depicted in Figure 3b, a sudden increase or decrease in peel force was evident, leading to the dissipation of release energy caused by the presence of micro-cracks during the peeling process. Consequently, despite cohesive failure, there was a lower level of reproducibility observed among the samples.

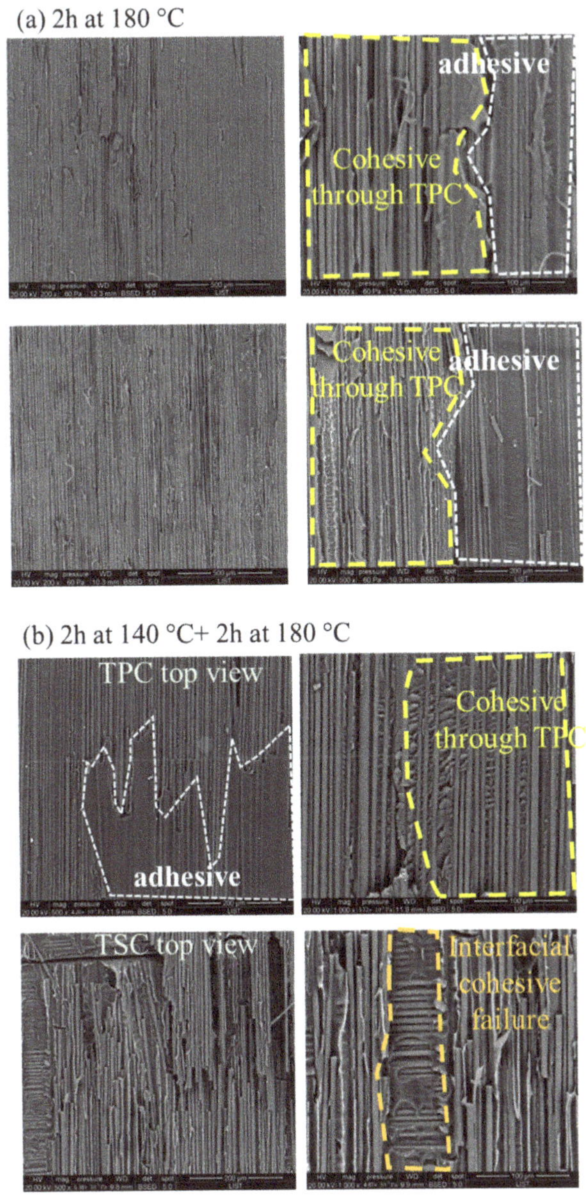

Figure 4. *Cont.*

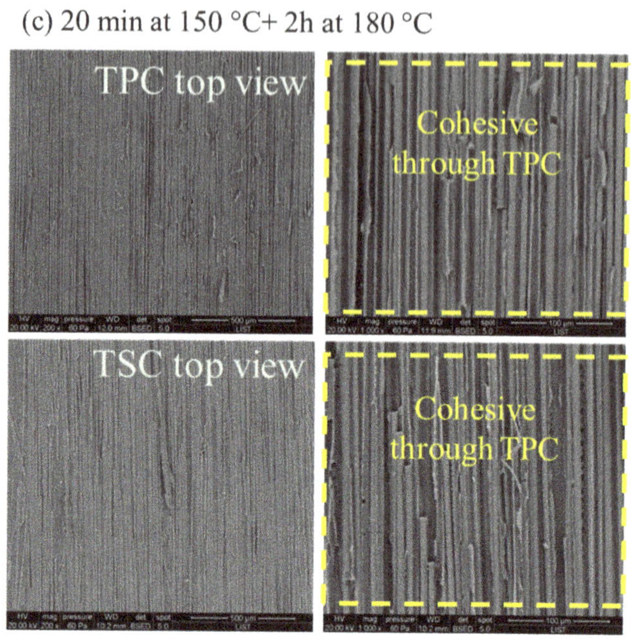

Figure 4. SEM images of the fracture surfaces of the co-cured samples after peel tests: (**a**) 2 h at 180 °C, (**b**) 2 h at 140 °C + 2 h at 180 °C, (**c**) 20 min at 150 °C + 2 h at 180 °C.

For the co-cured joint interface subjected to a curing cycle of 2 h at 150 °C followed by 2 h at 180 °C, the surface of the fracture exhibited greater roughness compared to the other samples cured under different cycles. Generally, a rougher fracture surface indicates increased material toughness. As depicted in Figure 5, we observe 'tied' fibres in this scenario, where strips of the matrix have been cleanly peeled from the fibre surface. These occurrences result from the fibres being pulled from the surfaces during the peeling test, consequently inducing localised shear at the interface. These localised deformations are well-correlated with changes in crack speed (stick–slip growth) or the partial arrest of the 10^{-3} crack front.

These findings offer exciting prospects for the assembly of hybrid structures through co-curing. However, the underlying physical mechanisms responsible for submicron interdiffusion need further exploration. The interdiffusion of resin molecules into thermoplastic materials is acknowledged to bring about alterations in the microstructure through processes like dissolution and swelling. This, in turn, leads to modifications in the physical and mechanical properties of the thermoplastics, influencing diffusion and interphase formation [50]. However, a comprehensive understanding of the interdiffusion process involving a reactive resin into thermoplastics during the co-curing process remains largely unexplored due to several challenges and complexities. These challenges include: (i) the continual changes in the molecular weight and structure of the thermosetting resin during the curing process; (ii) the substantial influence of curing reactions and interdiffusion on environmental conditions; and (iii) the unclear mechanism of diffusion and post-diffusion phenomena, such as gelation and phase separation [51]. Our ongoing experimental investigation of the interphase involves reproducing the diffusion phenomena without carbon fibre. This approach allows for more precise sample preparation conducive to sub-micron observations using atomic force microscopy (AFM).

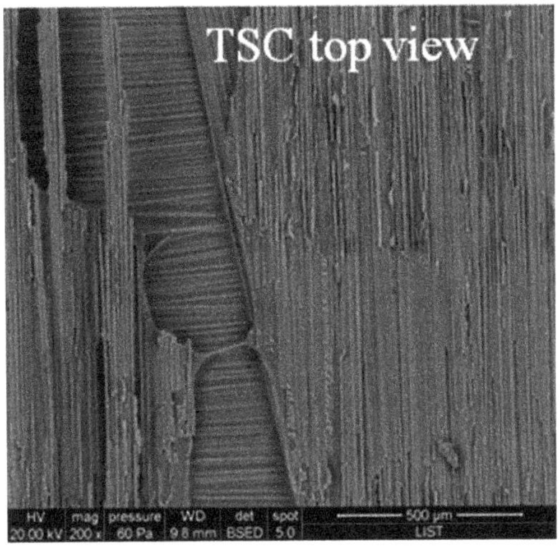

Figure 5. SEM images of the fracture surfaces of the co-cured samples at 150 °C for 2 h and at 180 °C for 2 h. TPS stands for thermoset.

4. Conclusions

This study introduces an innovative technique involving the co-cure bonding of dissimilar carbon fibre/epoxy (CF/Ep) and carbon poly-ether-ether-ketone (CF/PEEK) composites. An enhancement in mechanical interlocking was achieved by subjecting the C/PEEK tape surface to atmospheric plasma treatment. Key aspects of this approach include: (i) the implementation of a one-step co-cure bonding process that preserves the integrity of the thermoset components; and (ii) the simplicity of this welding process, obviating the need for an intermediary thermoplastic interlayer. The insights derived from this research showed that:

(1) The curing cycle exerts influential contributions to thebonding characterisation of co-cured interfaces of PEEK/RTM6. The results showed that curing temperature and time are of high importance, and hence the peeling test results do help in understanding which cure cycles can lead to strong bonds between PEEK/RTM6.
(2) While peeling force for the cure cycle of 2 h at 140 °C + 2 h at 180 °C showed higher values in some of the iterations, the curing cycle of 2 h at 150 °C + 2 h at 180 °C showed reproducible results, and the adhesion between PEEK and RTM6 tends to be more stable.

Further study would have to be made to measure the impact of peeling velocities where stick–slip behaviour becomes prominent. In forthcoming investigations, we plan to explore how surface roughness affects the specific characteristics of the stick–slip pattern.

Funding: This DefenceTech Project called INSHO (Innovative manufacturing approach for in-situ consolidation of high-performance thermoplastic composite sandwich structure based on honeycomb core) benefits from the shared financial support by the Ministry of Foreign and European Affairs, Directorate of Defence, the Ministry of Economy, and the Luxembourg National Research Fund (FNR) under the research grant DEFENCE22/MS/17800381.

Institutional Review Board Statement: Not applicable.

Informed Consent Statement: Not applicable.

Data Availability Statement: Data are contained within the article.

Acknowledgments: The authors greatly appreciate the support of Sébastien Gergen, Sébastien Klein and Loic Borghini for the plasma treatment, manufacturing of composite plates and mechanical tests from Luxembourg Institute of science and technology. This scientific paper is dedicated to the memory of our dear mentor and companion, Gérard Joly (1949–2023), whose remarkable inventions and contributions to the field of thermoplastic composite welding continue to inspire and guide our work.

Conflicts of Interest: The authors declare no conflict of interest.

References

1. Chawla, K.K. *Composite Materials*; Springer Science and Business Media LLC: Dordrecht, The Netherlands, 2019; ISBN 9783030289829.
2. Yousefpour, A.; Hojjati, M.; Immarigeon, J.-P. Fusion Bonding/Welding of Thermoplastic Composites. *J. Thermoplast. Compos. Mater.* **2004**, *17*, 303–341. [CrossRef]
3. Villegas, I.F.; van Moorleghem, R. Ultrasonic welding of carbon/epoxy and carbon/PEEK composites through a PEI thermoplastic coupling layer. *Compos. Part A Appl. Sci. Manuf.* **2018**, *109*, 75–83. [CrossRef]
4. Otheguy, M.; Gibson, A.; Robinson, A. Fusion bonding of structural T-joints for thermoplastic composite boats. *J. Thermoplast. Compos. Mater.* **2013**, *26*, 419–442. [CrossRef]
5. Ahmed, T.; Stavrov, D.; Bersee, H.; Beukers, A. Induction welding of thermoplastic composites—An overview. *Compos. Part A Appl. Sci. Manuf.* **2006**, *37*, 1638–1651. [CrossRef]
6. Panneerselvam, K.; Aravindan, S.; Haq, A.N. Study on resistance welding of glass fiber reinforced thermoplastic composites. *Mater. Des.* **2012**, *41*, 453–459. [CrossRef]
7. Stavrov, D.; Bersee, H. Resistance welding of thermoplastic composites-an overview. *Compos. Part A Appl. Sci. Manuf.* **2005**, *36*, 39–54. [CrossRef]
8. Van Ingen, J.W. Thermoplastic Orthogrid Fuselage Shell. *SAMPE* **2016**, *52*, 7–15. Available online: https://www.researchgate.net/publication/307926694 (accessed on 19 October 2023).
9. Ageorges, C.; Ye, L.; Hou, M. Advances in fusion bonding techniques for joining thermoplastic matrix composites: A review. *Compos. Part A Appl. Sci. Manuf.* **2001**, *32*, 839–857. [CrossRef]
10. Villegas, I.F.; Moser, L.; Yousefpour, A.; Mitschang, P.; Bersee, H.E. Process and performance evaluation of ultrasonic, induction and resistance welding of advanced thermoplastic composites. *J. Thermoplast. Compos. Mater.* **2013**, *26*, 1007–1024. [CrossRef]
11. Amend, P.; Frick, T.; Schmidt, M. Experimental Studies on Laser-based Hot-melt Bonding of thermosetting Composites and Thermoplastics. *Phys. Procedia* **2011**, *12*, 166–173. [CrossRef]
12. Ageorges, C.; Ye, L. Resistance welding of thermosetting composite/thermoplastic composite joints. *Compos. Part A Appl. Sci. Manuf.* **2001**, *32*, 1603–1612. [CrossRef]
13. Van Ingen, J.W.; Buitenhuis, A.; Van Wijngaarden, M.; Simmons, F. Development of the Gulfstream G650 induction welded thermoplastic elevators and rudder. In Proceedings of the International SAMPE Symposium and Exhibition, Seattle, WA, USA, 18–21 May 2010.
14. Deng, S.; Djukic, L.; Paton, R.; Ye, L. Thermoplastic–epoxy interactions and their potential applications in joining composite structures—A review. *Compos. Part A Appl. Sci. Manuf.* **2015**, *68*, 121–132. [CrossRef]
15. Yuan, X.; Jayaraman, K.; Bhattacharyya, D. Effects of plasma treatment in enhancing the performance of woodfibre-polypropylene composites. *Compos. Part A Appl. Sci. Manuf.* **2004**, *35*, 1363–1374. [CrossRef]
16. Perrin, H.; Mertz, G.; Senoussaoui, N.-L.; Borghini, L.; Klein, S.; Vaudemont, R. Surface functionalization of thermoset composite for infrared hybrid welding. *Funct. Compos. Mater.* **2021**, *2*, 10. [CrossRef]
17. Renart, J.; Costa, J.; Sarrado, C.; Budhe, S.; Turon, A.; Rodríguez-Bellido, A. Mode I fatigue behaviour and fracture of adhesively-bonded fibre-reinforced polymer (FRP) composite joints for structural repairs. In *Fatigue and Fracture of Adhesively-Bonded Composite Joints*; Elsevier: Amsterdam, The Netherlands, 2015; pp. 121–147. [CrossRef]
18. Wu, W.; Abliz, D.; Jiang, B.; Ziegmann, G.; Meiners, D. A novel process for cost effective manufacturing of fiber metal laminate with textile reinforced pCBT composites and aluminum alloy. *Compos. Struct.* **2014**, *108*, 172–180. [CrossRef]
19. Wu, W.; Klunker, F.; Xie, L.; Jiang, B.; Ziegmann, G. Simultaneous binding and ex situ toughening concept for textile reinforced pCBT composites: Influence of preforming binders on interlaminar fracture properties. *Compos. Part A Appl. Sci. Manuf.* **2013**, *53*, 190–203. [CrossRef]
20. Fu, X.; Wu, X.; Huang, G.; Li, W.; Kang, S.; Wang, L.; Luo, J.; Pan, Z.; Lu, W. Fusion Bonding Possibility for Incompatible Polymers by the Novel Ultrasonic Welding Technology: Effect of Interfacial Compatibilization. *ACS Omega* **2022**, *7*, 14797–14806. [CrossRef]
21. Moretti, L.; Olivier, P.; Castanié, B.; Bernhart, G. Experimental study and in-situ FBG monitoring of process-induced strains during autoclave co-curing, co-bonding and secondary bonding of composite laminates. *Compos. Part A Appl. Sci. Manuf.* **2021**, *142*, 106224. [CrossRef]
22. Hasan, Z.; Rader, J.; Olson, A.; Turpin, D.; Onge, R.S.; Amback, J. Design, analysis and fabrication of thick co-cured wing structures. *Compos. Part B Eng.* **2019**, *177*, 107335. [CrossRef]
23. Mohan, J.; Ivanković, A.; Murphy, N. Mixed-mode fracture toughness of co-cured and secondary bonded composite joints. *Eng. Fract. Mech.* **2015**, *134*, 148–167. [CrossRef]
24. Kim, G.-H.; Choi, J.-H.; Kweon, J.-H. Manufacture and performance evaluation of the composite hat-stiffened panel. *Compos. Struct.* **2010**, *92*, 2276–2284. [CrossRef]

25. Dhilipkumar, T.; Rajesh, M. Enhancing strength and stiffness of composite joint through co-cure technique. *Compos. Commun.* **2021**, *27*, 100878. [CrossRef]
26. Quan, D.; Zhao, G.; Scarselli, G.; Alderliesten, R. Co-curing bonding of carbon fibre/epoxy composite joints with excellent structure integrity using carbon fibre/PEEK tapes. *Compos. Sci. Technol.* **2022**, *227*, 109567. [CrossRef]
27. Shi, H.; Sinke, J.; Benedictus, R. Surface modification of PEEK by UV irradiation for direct co-curing with carbon fibre reinforced epoxy prepregs. *Int. J. Adhes. Adhes.* **2017**, *73*, 51–57. [CrossRef]
28. Kosmachev, P.; Alexenko, V.; Bochkareva, S.; Panin, S. Deformation Behavior and Fracture Patterns of Laminated PEEK- and PI-Based Composites with Various Carbon-Fiber Reinforcement. *Polymers* **2021**, *13*, 2268. [CrossRef]
29. Yao, C.; Qi, Z.; Chen, W.; Zhang, C. Experimental study on CF/PEEK thermoplastic fastener: Effects of fastener matrix crystallinity and fibre content on the strength of single-lap joint. *Compos. Part B Eng.* **2021**, *213*, 108737. [CrossRef]
30. Kosmachev, P.V.; Panin, S.V.; Panov, I.L.; Bochkareva, S.A. Surface Modification of Carbon Fibers by Low-Temperature Plasma with Runaway Electrons for Manufacturing PEEK-Based Laminates. *Materials* **2022**, *15*, 7625. [CrossRef]
31. Wu, Z.; Yi, X.-S.; Wilkinson, A. Interlaminar fracture toughness of carbon fibre/RTM6-2 composites toughened with thermoplastic-coated fabric reinforcement. *Compos. Part B Eng.* **2017**, *130*, 192–199. [CrossRef]
32. Zotti, A.; Elmahdy, A.; Zuppolini, S.; Borriello, A.; Verleysen, P.; Zarrelli, M. Aromatic Hyperbranched Polyester/RTM6 Epoxy Resin for EXTREME Dynamic Loading Aeronautical Applications. *Nanomaterials* **2020**, *10*, 188. [CrossRef]
33. *Product Data Sheet: HexTow® AS4C Carbon Fiber*; Hexcel: Stamford, CT, USA, 2023.
34. *APTIV ®2000 Series Films*; Victrex: Seoul, Republic of Korea. Available online: www.aptivfilms.com (accessed on 19 October 2023).
35. *HexFlow ® RTM6-2 Product Data 180 °C Bi-Component Epoxy System for Resin Transfer Moulding and In-Fusion Technologies*; Hexcel: Stamford, CT, USA, 2015.
36. Berthet, F. Comparaison des cinétiques de réaction de la résine RTM6 à l'aide des diagrammes TTT. *Rev. Des Compos. Matériaux Avancés* **2010**, *20*, 25. [CrossRef]
37. Navabpour, P.; Nesbitt, A.; Degamber, B.; Fernando, G.; Mann, T.; Day, R. Comparison of the curing kinetics of the RTM6 epoxy resin system using differential scanning calorimetry and a microwave-heated calorimeter. *J. Appl. Polym. Sci.* **2006**, *99*, 3658–3668. [CrossRef]
38. Lionetto, F.; Moscatello, A.; Maffezzoli, A. Effect of binder powders added to carbon fiber reinforcements on the chemoreology of an epoxy resin for composites. *Compos. Part B Eng.* **2017**, *112*, 243–250. [CrossRef]
39. Karkanas, P.I.; Partridge, I.K.; Attwood, D. Modelling the Cure of a Commercial Epoxy Resin for Applications in Resin Transfer Moulding. *Polym. Int.* **1996**, *41*, 183–191. [CrossRef]
40. *EN 28510-1:2014*; Adhesives—Peel Test for a Flexible-Bonded-to-Rigid Test Specimen Assembly—Part 1: 90° Peel. European Committee for Standardization: Brussels, Belgium, 2014.
41. Xu, D.-B.; Hui, C.-Y.; Kramer, E.J. Interface fracture and viscoelastic deformation in finite size specimens. *J. Appl. Phys.* **1992**, *72*, 3305–3316. [CrossRef]
42. Peng, Z.; Wang, C.; Chen, L.; Chen, S. Peeling behavior of a viscoelastic thin-film on a rigid substrate. *Int. J. Solids Struct.* **2014**, *51*, 4596–4603. [CrossRef]
43. Bartlett, M.D.; Case, S.W.; Kinloch, A.J.; Dillard, D.A. Peel tests for quantifying adhesion and toughness: A review. *Prog. Mater. Sci.* **2023**, *137*, 101086. [CrossRef]
44. Whitney, J.; Nuismer, R. Stress Fracture Criteria for Laminated Composites Containing Stress Concentrations. *J. Compos. Mater.* **1974**, *8*, 253–265. [CrossRef]
45. Blackman, B.; Kinloch, A.; Sanchez, F.R.; Teo, W.; Williams, J.; Blackman, B.; Kinloch, A.; Sanchez, F.R.; Teo, W.; Williams, J. The fracture behaviour of structural adhesives under high rates of testing. *Eng. Fract. Mech.* **2009**, *76*, 2868–2889. [CrossRef]
46. Satheesh, B.; Tonejc, M.; Potakowskyj, L.; Pletz, M.; Fauster, E.; Kaynak, B.; Schledjewski, R. Peel strength characterisation on ply/ply interface using wedge and T-peel/pull-type tests. *Polym. Polym. Compos.* **2018**, *26*, 431–445. [CrossRef]
47. Sacchetti, F.; Grouve, W.; Warnet, L.; Villegas, I.F. Woven fabric composites: Can we peel it? *Procedia Struct. Integr.* **2016**, *2*, 245–252. [CrossRef]
48. Cortet, P.-P.; Dalbe, M.-J.; Guerra, C.; Cohen, C.; Ciccotti, M.; Santucci, S.; Vanel, L. Intermittent stick-slip dynamics during the peeling of an adhesive tape from a roller. *Phys. Rev. E* **2013**, *87*, 022601. [CrossRef] [PubMed]
49. Cortet, P.-P.; Ciccotti, M.; Vanel, L. Imaging the stick–slip peeling of an adhesive tape under a constant load. *J. Stat. Mech. Theory Exp.* **2007**, *2007*, P03005. [CrossRef]
50. Pierce, F.; Perahia, D.; Grest, G.S. Interdiffusion of Short Chain Oligomers into an Entangled Polymer Film. *Macromolecules* **2009**, *42*, 7969–7973. [CrossRef]
51. Zanjani, J.S.M.; Baran, I.; Akkerman, R. Characterization of interdiffusion mechanisms during co-bonding of unsaturated polyester resin to thermoplastics with different thermodynamic affinities. *Polymer* **2020**, *209*, 122991. [CrossRef]

Disclaimer/Publisher's Note: The statements, opinions and data contained in all publications are solely those of the individual author(s) and contributor(s) and not of MDPI and/or the editor(s). MDPI and/or the editor(s) disclaim responsibility for any injury to people or property resulting from any ideas, methods, instructions or products referred to in the content.

MDPI
St. Alban-Anlage 66
4052 Basel
Switzerland
www.mdpi.com

Materials Editorial Office
E-mail: materials@mdpi.com
www.mdpi.com/journal/materials

Disclaimer/Publisher's Note: The statements, opinions and data contained in all publications are solely those of the individual author(s) and contributor(s) and not of MDPI and/or the editor(s). MDPI and/or the editor(s) disclaim responsibility for any injury to people or property resulting from any ideas, methods, instructions or products referred to in the content.

www.ingramcontent.com/pod-product-compliance
Lightning Source LLC
Chambersburg PA
CBHW040716270926
51850CB00023B/2335
9783036599632